Exploiting the Power of Group Differences

Using Patterns to Solve Data Analysis Problems

Synthesis Lectures on Data Mining and Knowledge Discovery

Editors
Jiawei Han, *University of Illinois at Urbana-Champaign*
Lise Getoor, *University of California, Santa Cruz*
Wei Wang, *University of California, Los Angeles*
Johannes Gehrke, *Cornell University*
Robert Grossman, *University of Chicago*

Synthesis Lectures on Data Mining and Knowledge Discovery is edited by Jiawei Han, Lise Getoor, Wei Wang, Johannes Gehrke, and Robert Grossman. The series publishes 50- to 150-page publications on topics pertaining to data mining, web mining, text mining, and knowledge discovery, including tutorials and case studies. Potential topics include: data mining algorithms, innovative data mining applications, data mining systems, mining text, web and semi-structured data, high performance and parallel/distributed data mining, data mining standards, data mining and knowledge discovery framework and process, data mining foundations, mining data streams and sensor data, mining multi-media data, mining social networks and graph data, mining spatial and temporal data, pre-processing and post-processing in data mining, robust and scalable statistical methods, security, privacy, and adversarial data mining, visual data mining, visual analytics, and data visualization.

Exploiting the Power of Group Differences

Using Patterns to Solve Data Analysis Problems

Synthesis Lectures on Data Mining and Knowledge Discovery

Editors

Jiawei Han, *University of Illinois at Urbana-Champaign*
Lise Getoor, *University of California, Santa Cruz*
Wei Wang, *University of California, Los Angeles*
Johannes Gehrke, *Cornell University*
Robert Grossman, *University of Chicago*

Synthesis Lectures on Data Mining and Knowledge Discovery is edited by Jiawei Han, Lise Getoor, Wei Wang, Johannes Gehrke, and Robert Grossman. The series publishes 50- to 150-page publications on topics pertaining to data mining, web mining, text mining, and knowledge discovery, including tutorials and case studies. Potential topics include: data mining algorithms, innovative data mining applications, data mining systems, mining text, web and semi-structured data, high performance and parallel/distributed data mining, data mining standards, data mining and knowledge discovery framework and process, data mining foundations, mining data streams and sensor data, mining multi-media data, mining social networks and graph data, mining spatial and temporal data, pre-processing and post-processing in data mining, robust and scalable statistical methods, security, privacy, and adversarial data mining, visual data mining, visual analytics, and data visualization.

Exploiting the Power of Group Differences: Using Patterns to Solve Data Analysis Problems
Guozhu Dong

ISBN: 978-3-031-00785-9   paperback
ISBN: 978-3-031-01935-5   ebook
ISBN: 978-3-031-00108-6   hardcover

DOI: 10.1007/978-3-031-01935-5

A Publication in the Springer series
SYNTHESIS LECTURES ON DATA MINING AND KNOWLEDGE DISCOVERY

Lecture #16
Series Editors: Jiawei Han, *University of Illinois at Urbana-Champaign*
                Lise Getoor, *University of California, Santa Cruz*
                Wei Wang, *University of California, Los Angeles*
                Johannes Gehrke, *Cornell University*
                Robert Grossman, *University of Chicago*
Series ISSN
Print 2151-0067   Electronic 2151-0075

Exploiting the Power of Group Differences

Using Patterns to Solve Data Analysis Problems

Guozhu Dong
Wright State University

SYNTHESIS LECTURES ON DATA MINING AND KNOWLEDGE DISCOVERY #16

ABSTRACT

This book presents pattern-based problem-solving methods for a variety of machine learning and data analysis problems. The methods are all based on techniques that exploit the power of group differences. They make use of group differences represented using emerging patterns (aka contrast patterns), which are patterns that match significantly different numbers of instances in different data groups. A large number of applications outside of the computing discipline are also included.

Emerging patterns (EPs) are useful in many ways. EPs can be used as features, as simple classifiers, as subpopulation signatures/characterizations, and as triggering conditions for alerts. EPs can be used in gene ranking for complex diseases since they capture multi-factor interactions. The length of EPs can be used to detect anomalies, outliers, and novelties. Emerging/contrast pattern-based methods for clustering analysis and outlier detection do not need distance metrics, avoiding pitfalls of the latter in exploratory analysis of high dimensional data. EP-based classifiers can achieve good accuracy even when the training datasets are tiny, making them useful for exploratory compound selection in drug design. EPs can serve as opportunities in opportunity-focused boosting and are useful for constructing powerful conditional ensembles. EP-based methods often produce interpretable models and results. In general, EPs are useful for classification, clustering, outlier detection, gene ranking for complex diseases, prediction model analysis and improvement, and so on.

EPs are useful for many tasks because they represent group differences, which have extraordinary power. Moreover, EPs represent multi-factor interactions, whose effective handling is of vital importance and is a major challenge in many disciplines.

Based on the results presented in this book, one can clearly say that patterns are useful, especially when they are linked to issues of interest.

We believe that many effective ways to exploit group differences' power still remain to be discovered. Hopefully this book will inspire readers to discover such new ways, besides showing them existing ways, to solve various challenging problems.

KEYWORDS

data mining, machine learning, data analytic, classification, regression, clustering, anomaly detection, outlier detection, intrusion detection, compound selection, complex disease analysis, extreme instance selection, factor ranking, prediction model analysis, group difference analysis, feature, multifactor interaction, diverse relationship, heterogeneity, boosting, ensemble, association rule, emerging pattern, contrast pattern, frequent pattern, distance metric, interpretability

To my wife for her love and support!

– Guozhu Dong

Contents

Acknowledgments

Many researchers directly contributed to results reported in this book, concerning (1) the development of computational methods that use emerging/contrast patterns to tackle various machine learning and data mining/analysis problems, and (2) the utilization of emerging/contrast patterns and methods to solve challenging problems outside of the computing discipline. Their work demonstrated the power of group differences and emerging/contrast patterns. Many other researchers indirectly contributed to results reported in this book through results on the mining of emerging/contrast patterns. I am grateful to all of those researchers.

I appreciate the helpful comments as well as useful references provided by James Bailey, Bruno Crémilleux, Ramamohanarao (Rao) Kotagiri, Jinyan Li, José Francisco Martínez-Trinidad, and Limsoon Wong. They helped improve the quality and completeness of this book.

I am grateful to Jiawei Han for helpful and encouraging suggestions, and to Diane D. Cerra and C.L. Tondo for assistance on various technical issues related to the preparation of this manuscript. I also appreciate the helpful comments from the reviewers, which helped improve the book.

Guozhu Dong
February 2019

CHAPTER 1

Introduction and Overview

This chapter starts with a discussion on the importance of group differences. It then provides brief summaries of each of the other 10 chapters of this book, followed by a high level summary of known uses of group differences and emerging patterns. Then it gives a brief summary of unique properties of emerging pattern based methods. It concludes with a brief overview of related topics not covered in this book.

As discussed in the abstract, this book focuses on pattern-based problem-solving techniques that exploit the power of group differences. The techniques solve problems in machine learning, data mining, and data analytics. As will be seen in later chapters, the techniques have been found useful in a variety of other disciplines that rely on data-driven problem-solving methods.

The techniques mostly make use of group differences represented by emerging patterns, namely patterns whose supports in two data groups are highly different.

> *A data group is a subset of a given dataset/application under study, often defined by classes or conditions on data instances.*

More details on the basics of data groups are given in Chapter 3.

This book does not discuss the majority of the emerging pattern mining algorithms that have been presented in the literature since it is focused on emerging-pattern based problem-solving techniques. However, it includes a simple and useful mining method that is flexible and easy to use.

We note that emerging pattern mining is a special approach for contrast mining and supervised descriptive pattern mining. Contrast mining is concerned with the mining of patterns and models that distinguish different data groups.

1.1 IMPORTANCE OF GROUP DIFFERENCES

Analyzing and making use of group difference is important in a variety of human activities, ranging from everyday thinking and decision making to scientific investigation and complex problem solving. The discussion below illustrates, using a small sample of such activities.

Dividing objects (which can be people, things, or situations) into multiple categories and using different ways to treat objects in different categories is a fundamental and frequently used approach by humans. For example, we divide people into age groups, patients into different disease classes, and situations into "dangerous" and "normal." Such division makes it easier to organize/learn the underlying knowledge accumulated over time, to communicate between humans, and so on. Sometimes such division can even help humans live better/safer, and help them improve their survival in the natural world, as illustrated by the use of the mechanisms associated with "fear" and "disgust" [49, 142, 154].

Many concepts of broad interest, including contrast, change, trend, characteristics, uniqueness, heterogeneity, and diversity, are closely related to differences between/among data groups (explicit or implicit) of some sorts. Keeping up with the changes is interesting as it is important to individuals' future. Knowing the characteristics and uniqueness of vital (data) groups, for example, the positive class of a serious disease, plays a key role in the understanding and recognition of objects of such groups. Having a sound understanding of the heterogeneity and diversity of a given population, which is about the existence of different groups with highly distinct properties and behaviors, is the starting point of treating different groups in the best ways possible.

Group difference is the foundation of several traditional statistical inference approaches. For example, the well known t test allows one to draw conclusions based on the difference in the mean values of an attribute in two data groups. Establishing the existence of significant difference between two given groups is sometimes used as the basis for major conclusions such as "the drug is effective." The fact that researchers in statistics have promoted going beyond "difference in the means" [175] also indicates that group difference is and continues to be an important issue.

The importance of analyzing and utilizing group difference has also been widely recognized in data mining and machine learning. Several pattern/rule types linked to group differences have been proposed and many papers have been published on them. These include classification rules (e.g., [167]), emerging patterns [56], contrast sets [25], and subgroups having highly unique properties [186, 217]. Several other names have been used as synonyms of patterns capturing group differences, including contrast patterns, discriminating patterns, and distinguishing patterns.

A large number of research papers in diverse scientific disciplines have used group differences to derive significant findings or methods. The subsequent chapters will refer to many such papers. The tasks considered, as well the roles played by group differences, in those papers, can be found in Figures 1.1 and 1.2. Representative papers not cited in later chapters are discussed in Section 1.6 below.

1.2   SUMMARY OF CHAPTERS

Among the 11 chapters in this book, 8 discuss emerging pattern (EP) based techniques. Besides this chapter, there is a chapter on general preliminaries and another on the basics of EPs. This section gives brief summaries of Chapters 2–11.

Chapter 2 presents the following preliminary concepts and methods that are needed in multiple later chapters of this book: attributes, features, data instances, tuples, classes, data types, binning, patterns, matching datasets, support, frequent patterns, equivalence classes of patterns, closed patterns, minimal generators, and borders. Other preliminaries will be given in the chapters as needed.

Chapter 3 starts with the basics for emerging patterns. It then presents a simple algorithm for mining emerging patterns, which can even be applied by hand on small datasets. Then it discusses what emerging patterns can represent with respect to the analysis of group differences, and it points to parts of the book that illustrate various kinds of uses of emerging patterns. It also discusses traditional ways to analyze group differences. It concludes with a discussion of related issues.

Chapter 4 focuses on a "discriminating pattern aggregation-based approach," called CAEP, for classification. After some background materials on classification, it presents the CAEP approach. It then gives a discussion on CAEP's performance in experiments and applications, as well as its uniqueness and strengths. Then it presents a lazy instance-based approach, called DeEPs, which also follows the aggregation approach. It then discusses CAEP's relationship with other rule/pattern-based classifiers.

Chapter 5 first uses experiments to show that CAEP has good performance and outperforms other classification algorithms, when the training datasets are tiny. It then presents an iterative algorithm for compound selection, which minimizes the amount of human effort in the selection process. The algorithm exploits CAEP's ability to perform well on tiny training data. This chapter also generalizes the compound selection problem to the semi-supervised extreme instance selection problem, and gives an algorithm to solve the problem by adapting the iterative algorithms for compound selection. It also compares the semi-supervised extreme instance selection problem with semi-supervised learning.

Chapter 6 presents an emerging pattern based method for intrusion detection and outlier detection called OCLEP, which is based on the use of the length of jumping emerging patterns. Key properties of OCLEP include the following: (1) It uses only *one-class training data* to build an intrusion detection system. (2) The constructed intrusion detection system does not rely on a mathematical model, making it hard for attackers to figure out the details of the detection system. (3) The constructed intrusion detection system does not use distance metrics, and the method can effectively handle both categorical attributes and numerical attributes. (4) The method has strength in interpretability for analyzing discovered intruders (and outliers when used for outlier detection). (5) It is a lazy instance-based approach. OCLEP can be used to perform outlier detection when the problem is treated as a one-class training problem.

Chapter 7 presents a contrast pattern based method called CPCQ for clustering quality evaluation. CPCQ does not use distance metrics. The chapter starts with some background materials on clustering quality evaluation. Then it gives the rationale for CPCQ, followed by techniques to measure the quality of individual contrast patterns (CPs) and to measure the diversity of (sets of) high-quality CPs. Then it defines the CPCQ measure and presents a method to compute some best groups of CPs to maximize CPCQ values. This is followed by discussion on experimental evaluation of CPCQ. Finally, some concluding remarks are given.

Chapter 8 presents a pattern-based clustering algorithm called CPC, together with experiments on textual blog data, to illustrate CPC and its strength. CPC works with frequent patterns mined from given data to produce clusterings aimed at maximizing CPCQ. So CPC is aimed at maximizing the quality and diversities of contrast patterns that distinguish the produced clusters. Special strengths of CPC include: (1) CPC does not use distance metrics and (2) the method produces interpretable results by associating each cluster with a set of contrast patterns characterizing it.

Chapters 6, 7, and 8 demonstrate that characteristics of emerging/contrast patterns can be used to indicate qualities of data mining and machine learning results.

Chapter 9 considers the challenging problem of gene ranking for complex diseases when the number of genes is in the thousands. It presents an emerging pattern based method for this problem called IBIG (Interaction-Based Importance of Genes). IBIG ranks genes for complex diseases by taking multi-gene interactions into consideration, and a gene's rank given by IBIG is based on the number of influential interactions that the gene participates in. IBIG uses jumping emerging patterns (JEPs) to represent multi-gene interactions. Given a set of mined JEPs, a gene's IBIG rank depends on the number of high-quality JEPs involving the gene and the quality of those JEPs; a gene is ranked high if it occurs in many high-quality JEPs. The *quality* of a JEP is *measured by its support in its home class*. In order to get the optimal ranking of genes, some optimal set of JEPs, which may involve any of the thousands of genes, needs to be mined. IBIG does that by using an iterative algorithm together with a so-called gene-club technique, which are especially designed to address the challenge posed by the presence of thousands of genes. The IBIG method often mines high-quality JEPs that have much larger home-class support than other methods, in addition to producing the IBIG gene ranking.

Chapter 10 presents pattern aided prediction (PXP) models, and the CPXP approach, which uses contrast patterns to construct PXP models. Here, prediction covers both regression and classification. Moreover, patterns are used as conditions to define and characterize subpopulations whose local prediction models are substantially different (a) from the global model on the whole population (covering all data) and (b) perhaps also from local models for other subpopulations. Different from traditional boosting, the CPXP approach uses both opportunity and reward to guide the boosting process. Different from traditional ensembles, the PXP models are conditional ensembles, as each member model is only applied to instances matching its associated pattern. The chapter also presents a so-called diverse predictor-response relationship concept. The concept of subpopulationwise conditional correlation analysis, involving a special adaptation of PXP and CPXP, is also discussed. Experimental results are summarized and discussed, including practical applications to water content prediction for soil and medical risk prediction for both traumatic brain injury and heart failure.

There are many interesting approaches and applications that involve the use of emerging/contrast patterns, and they are often quite different from each other. Chapter 11 gives an overview of representatives of these approaches and applications; the presentation is not very detailed due to space limit. From a technique-focused perspective, the sections of this chapter belong to six groups: Group 1 discusses representative approaches that use emerging/contrast patterns as group signatures/characterizations. They use such patterns to analyze given data groups, and perhaps also to solve other domain-specific tasks. Group 2 is concerned with representative applications that use emerging patterns as features. Group 3 is about using CAEP to solve a range of classification/recognition problems. Group 4 is about various ways to use emerging patterns to build classifiers. Group 5 is about using emerging patterns for classification over streaming data. Group 6 is about other applications. The chapter also contains a section that provides a high-level view from a "discipline and direction" focused perspective.

1.2.1   READING ORDER OF THE CHAPTERS
The dependency among the chapters depends on the background of the reader.

- For readers with basic knowledge of data mining and machine learning, the chapters can be read in any order.

- For other readers, Chapter 2 should be read first. After that, the chapters can be read in any order, except that Chapter 4 should be read before Chapter 5, and Chapter 7 should be read before Chapter 8.

1.3   KNOWN USES OF GROUP DIFFERENCES VIA EMERGING PATTERNS

Figure 1.1 gives a partial list of known roles of group differences and emerging patterns for a range of machine learning, data mining, and data analytic tasks; the inner six phrases are the roles, and the other phrases are tasks. Figure 1.2 gives a partial list of known application areas of group differences and emerging patterns. Some of the uses are described in detail in the subsequent chapters; others are described in less detail, mostly in Chapter 11.

The known uses can be described using four dimensions: generic task, role/function, application-specific task, and application domain.

- Values for the "generic tasks" dimension include: classification, regression, clustering, outlier detection, characterization, signature discovery, interaction discovery, interaction analysis, correlation analysis, trend analysis, model analysis, model boosting, and model quality evaluation. "Prediction" refers to "classification or regression or outlier detection." "Model" refers to "prediction model or clustering result." "Analysis" refers to "characterization or signature analysis or interaction analysis." The term "signature" includes artifacts such as "biomarker."

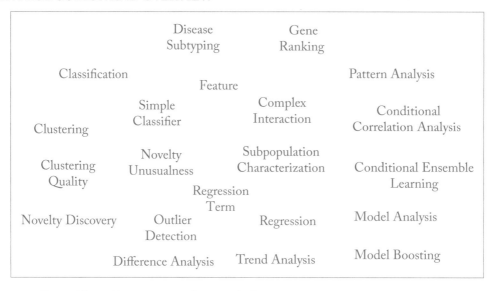

Figure 1.1: Partial list of known uses of group differences and emerging patterns: roles and tasks.

- Values for the "roles/functions" dimension include: feature, basic classifier, quality indicator, unusualness indicator, and subpopulation characterizer. Feature can be used for any of the four tasks of classification, regression, clustering, and outlier detection.

The values for the dimensions of application-specific tasks and application domains are application dependent and can vary a lot.

1.4   UNIQUE PROPERTIES OF EMERGING PATTERN BASED METHODS

Some emerging pattern (EP) based methods have unique properties. These include the following:

- EP-based methods for clustering analysis and outlier detection do not need distance metrics. This is a useful strength as the reliability of distance metrics is problematic in *exploratory analysis* of *high dimensional data*.

- EP-based classifiers can achieve good accuracy even when the training datasets are tiny, a property making them highly useful for exploratory compound selection in drug design and extreme instance selection.

- EPs can characterize subpopulations of data where a given prediction model makes large prediction errors, which can be used as opportunities in opportunity-focused boosting.

Toxicity Alert	Music Family Characterization	Disease Characterization and Diagnosis
Drug Design Compound Selection	Image/Video Activity Recognition	Unusual Patient Subtype Discovery
Street Crime Analysis	Complex Disease Gene Ranking	Short Text Intent Classification
Birth Defect Analysis		
Complex Disease Interaction Analysis	Inhibitor Prediction	Traumatic Brain Injury Analysis
Medical Operation Monitoring and Feedback	Customer Interest Discovery	Heart Failure Mortality Analysis
Biomarker Discovery	Adverse Drug Reaction Modeling	Structure Activity Relationship Analysis

Figure 1.2: Partial list of known uses of group differences and emerging patterns: application areas.

- EP-based methods for various data analytic tasks often produce interpretable models and results.

1.5 SCENARIOS WHERE EMERGING PATTERNS ARE ESPECIALLY USEFUL

As will be seen in later chapters, there are many situations where emerging patterns have been found useful. We now discuss the scenarios where emerging patterns are especially useful. In general, emerging patterns are especially useful for applications where

(1) all length-1 and length-2 patterns[1] have small growthRate, especially when their growthRate is close to 1, and,

(2) there are some emerging patterns with large growthRate and reasonable support.

This is because humans have been accustomed to making use of length-1 and length-2 patterns to solve problems, a skill humans learned through many generations. Observe that length-1 and length-2 patterns are not only easy to discover by observation without the help of computers, they are also easy to use. If all length-1 and length-2 patterns have small growthRate, then the underlying problem is not easy to solve using length-1 and length-2 patterns. So the problem

---

[1]The length of a pattern is the number of items (or single-attribute conditions) in it. The growthRate of a pattern is roughly its support ratio (see Chapter 3).

can be still very challenging to humans and may still be unsolved. If there exist emerging patterns with large growthRate, then these emerging patterns can be used as handles to complex interactions among attributes and can lead to a much better solution not available when limited to length-1 and length-2 patterns.

The author's experience with working with emerging patterns shows that emerging patterns with large growthRate and reasonable support often have length 4 or larger. Some examples of such emerging patterns are given in Chapter 9.

1.6   RELATED TOPICS NOT COVERED IN THIS BOOK

Representative applications not cited in later chapters include the following. Reference [187] used contrast mining to discover novel gene associations specific to autism subgroups. References [206] used contrast mining to help visualize and analyze spatiotemporal data about crimes. References [128] discussed the use of discriminative pattern mining in bioinformatics. Reference [26, 66] used discriminative patterns to analyze prediction model errors. Reference [179] introduced a method to select a concise set of discriminative patterns and then construct a generalized linear regression model using the selected patterns as features.

Representative papers on the mining of patterns linked to group differences include the following. References [25, 27] studied the mining of "contrast sets," defined to be patterns with large support differences between data groups. Reference [212] studied the problem of detecting differences between groups. Several papers [103, 218] studied the subgroup discovery problem, whose aim is to identify patterns whose matching datasets have unusual statistical (distributional) characteristics with respect to a special target variable (e.g., a class or a numerical response variable). Reference [44] studied direct discriminative pattern mining for effective classification. Reference [130] considered fast mining of high dimensional expressive contrast patterns using zero-suppressed binary decision diagrams. Reference [74] studied mining low-support discriminative patterns from dense and high-dimensional data. Reference [188] studied using constraints to generate and explore higher-order discriminative patterns. All these show that many researchers view mining patterns that are related to group differences as an important direction.

Several papers/books provided unifying surveys of the mining of contrast sets, emerging patterns, and subgroup discovery, including [153], [93], [35], [85], and [202]. Some of these also cover successful applications of the presented algorithms. For example, [202] includes a chapter on successful applications of "supervised descriptive pattern mining" in medicine, education, sociology, economy and energy, and so on.

Several papers explored the relationship of emerging patterns with other concepts: reference [196] examined the relation between jumping emerging patterns and rough set theory. Reference [17] linked FCA (formal concept analysis) and emerging patterns. Reference [76] used emerging pattern mining for exceptional pattern discovery.

CHAPTER 2

General Preliminaries

This chapter presents general preliminary concepts and methods concerning attributes, attribute binning, patterns, and pattern representation; they are needed in many chapters of this book. Specific preliminaries will be given in the chapters where they are needed. Key concepts discussed in this chapter are illustrated in Section 2.6.

2.1   ATTRIBUTES, FEATURES, AND VARIABLES

Attributes, also called *features* and *variables*, are used to describe certain properties of individual data objects. Examples include *age* and *major* for students. For a given application (at a fixed point in time), often a fixed set of attributes is implicitly chosen to describe all underlying data objects; each object takes a particular value for each relevant attribute. This results in a vector-based representation of the data objects.

Attributes fall into several types, including categorical, ordinal, and numerical. Different attribute types require different kinds of treatment, due to structural differences in their domains. (a) The domain of a categorical attribute is a set of discrete values. For example, *color* is a categorical attribute whose domain is $\{black, blue, brown, green, red, white, yellow\}$. (b) A special type of categorical attributes is binary, whose domain has exactly two values. (c) The domain of an ordinal attribute is a set of ordered values. The *degree* attribute is an ordinal attribute whose domain is $\{Bachelor, Master, Doctoral\}$; the three values are ordered in the listed order. (d) The domain of a numerical attribute is a set of numerical values. A numerical attribute is also called quantitative or continuous. For example, the *age* attribute is a numerical one whose domain is the set of integers between 0 and 150.

2.2   DATA INSTANCES AND DATASETS

Data instances are often described using either itemset-based representation or tuple-based representation. In the itemset-based representation, there is a fixed set of possible items, and a data instance is a set of items; such an instance is often called an itemset or a transaction. In the tuple-based representation, there is a fixed set of possible attributes, and a data instance is an ordered list of element values one for each attribute; such an instance is often called a tuple or a vector.

A dataset is simply a set of data instances.

The number of elements in a tuple is the tuple's arity. The tuples in a given dataset often have a fixed arity, which corresponds to the number of attributes. In the itemset-based representation, different instances may have different numbers of items, and the items (values) of different instances at a fixed position are not required to be associated with a fixed attribute. The itemset-based representation is often referred to as transactional data. The tuple-based representation is often referred to as relational data.

2.3 ATTRIBUTE BINNING AND DISCRETIZATION

For machine learning and data mining/analytics, there is often a need to perform binning or discretization on numerical attributes. Binning partitions the range of a numerical attribute into several intervals which are also called bins; for example, the range $[0, 30]$ is partitioned into these three bins: $[0, 10)$, $[10, 20)$, $[20, 30]$. The "[" and "]" symbols indicate that the corresponding end of the interval is closed—the corresponding bound is part of the interval. The "(" and ")" symbols indicate that the corresponding end of the interval is open—the corresponding bound is not part of the interval.

The result of a binning operation can be used to produce a coarse representation of data by mapping original values into the intervals they fall into. Moreover, the bin boundaries are often used as constants in single-attribute conditions (which are often called items) for use in patterns. We work often this way in this book.

Binning can be top-down, which splits given intervals into sub-intervals, or bottom-up, which merges smaller intervals into larger intervals. Binning can also be static, where binning happens just once before running other algorithms, or dynamic, where binning is performed inside another algorithm (perhaps multiple times). Static methods are often used; we focus on static ones in this book.

Many binning methods have been developed. They can be divided into two categories: a binning method is called *supervised* if the tuples have assigned classes and the method uses the class information, and it is called *unsupervised* otherwise [65]. Unsupervised binning methods include equi-width and equi-density. Supervised binning methods include the entropy-based method and the Class Distribution Curve-based binning (CDC) method. All four methods are described below.

Let A be a numerical attribute in a given dataset D, and let k be the desired number of discretized intervals for A. Let $[a_{\min}, a_{\max}]$ denote the *active range* of A, where $a_{\min}$ and $a_{\max}$ are respectively the minimum and maximum values of A in D.

The *equi-width* method divides A's active range into intervals of equal width. Specifically, the method uses the following intervals for A: $[a_{\min}, a_{\min} + w_e)$, $(a_{\min} + w_e, a_{\min} + 2w_e]$, ..., $(a_{\min} + (k-1)w_e, a_{\max}]$, where $w_e = \frac{a_{\max} - a_{\min}}{k}$.

For the following binning methods, it is customary to only use the mid-points of distinct consecutive values (in the sorted list of the values of the attribute) as potential bin boundaries to split bins. This makes the bins more "generalizable" on future data.

The *equi-density* method divides A's active range into intervals all having (as close to as possible) the same number of A values (including duplicates). Specifically, the method uses the intervals $[a_{\min}, a_1], (a_1, a_2], \ldots, (a_{k-1}, a_{\max}]$ such that the interval densities, $|\{t \mid t \in D, t(A) \in$ the ith interval$\}|$, are as close to $\frac{|D|}{k}$ as possible.

The entropy-based binning relies on the entropy measure. Let D' be a dataset having κ classes $C_1, \ldots, C_\kappa$. Let $p_i = \frac{|C_i|}{|D'|}$ for each i. The *entropy* [180] of D' is defined by

$$\text{entropy}(D') = -\sum_{i=1}^{\kappa} p_i \, \log_2 \, p_i.$$

An entropy value is often viewed as an indication of the purity of D'—the smaller the entropy value, the "purer" (or more "skewed") D' is.

The *entropy-based binning* method iteratively splits an interval into two intervals, starting by splitting the active range of A over the complete dataset. To determine the split value in some dataset D, the method [77] first sorts the A values in D into an increasing list $a_1, \ldots, a_n$. Then each mid-point between two distinct consecutive A values in the list is a candidate split value. Each split value v divides D into two subsets, $D_1 = \{t \in D \mid t(A) \leq v\}$ and $D_2 = \{t \in D \mid t(A) > v\}$. The *information gain* of a split v is defined to be

$$\text{IG}(v) = \text{entropy}(D) - \sum_{i=1}^{2} \frac{|D_i|}{|D|} \text{entropy}(D_i).$$

The split value v' that maximizes $\text{IG}(v)$ is chosen as the split value for A over D. This splits the interval of A into two intervals. If more intervals are needed, this method is used to find the best split value for A for one of the intervals (described by D_1) and the best split value for A for the other interval (described by D_2); then the better one is selected, leading to one more interval. This process is repeated until some stopping condition is satisfied.

Reference [183] introduced a supervised binning algorithm called CDC Binning. The algorithm uses a purity[1] measure, on the distribution of the classes in local neighborhoods (specified by a window width) of attribute values, to select bin boundaries. This approach allows one to separate the range of an attribute into several intervals so that (1) values in each interval all have a similar degree of purity and (2) values in different intervals have considerably different degrees of purity.

To ensure that a binning method applies to all unseen future data, after binning, one often replaces $a_{\min}$ by $-\infty$, and replaces $a_{\max}$ by $+\infty$.

---

[1]Purity in a neighborhood can be measured using entropy for the neighborhood, or ratio of counts for different classes in the neighborhood.

2.4   PATTERNS, MATCHING DATASETS, SUPPORTS, AND FREQUENT PATTERNS

In the most general form, a *pattern* is a condition, defined by a mathematical formula, on individual data instances that evaluates to either *true* or *false* on each data instance. Not all conditions on data instances are of interest as patterns. Only succinct conditions that are typically much simpler and smaller than the datasets they describe/match can be used as patterns. Moreover, computational complexity associated with pattern mining also leads to restrictions on conditions for use as patterns: complex conditions that are too hard to mine are often excluded as patterns.

There are many possible pattern definition languages. We discuss two frequently used ones below.

Patterns for transactional data are often given as itemsets. An *itemset* is a finite set of items. A transaction t is said to *satisfy* or *match* a pattern X if $X \subseteq t$ is true. It is customary to use shorthand when writing itemsets by omitting the brackets and commas whenever no confusions arise. For example, with $0, 1, \ldots, 9$ as items, 026 represents $\{0, 2, 6\}$.

Patterns for relational data are conjunctions of single-attribute conditions; single-attribute conditions have the form $A = a$ or $A \in \eta$, where a is a constant and η is an interval; $A \in \eta$ can be rewritten, for example, as $a' \leq A < a''$ if $\eta = [a', a'')$. It is easy to see that such patterns are simply the itemset-based patterns for the discretized data of the original relational data. The *satisfaction* of a single-attribute condition $A = a$ or $A \in \eta$ by a tuple t is defined in the natural manner. A tuple t *satisfies* a pattern X if each single-attribute condition in X is satisfied by t. Equivalently, we say that t *satisfies* an itemset X if the discretized version of t satisfies X in the transactional sense. The word "matches" is often used as a synonym of "satisfies."

Since the set of single-attribute conditions in a pattern is interpreted in the conjunction sense, we often write a pattern as the conjunction (AND) of the single-attribute conditions, with "AND" written as "&".

The *matching data* or *matching dataset* of a pattern X in a dataset D is given by $\text{mds}(X, D) = \{t \in D \mid t \text{ satisfies } X\}$. The *count* and *support* of X in D are given by $\text{count}(X, D) = |\text{mds}(X, D)|$ and $\text{supp}(X, D) = \frac{\text{count}(X, D)}{|D|}$. The concepts of itemset, count, and support given here are the same as in association mining [5]. A pattern X is *frequent* in D for a given support threshold σ if $\text{supp}(X, D) \geq \sigma$.

Given two patterns P and Q, P is a *sub-pattern* of Q, and Q is a *super-pattern* of P, if P is a subset of Q. Sub-patterns are more general than their super-patterns, as all instances matching the super-patterns match the sub-patterns. Moreover, sub-patterns are shorter than the super-patterns. We often prefer to have the minimal patterns (among a set of patterns having certain properties) in the set containment sense. Equivalently we prefer to have the patterns with a given property that have no proper sub-patterns also having the property.

2.5 EQUIVALENCE CLASSES, CLOSED PATTERNS, MINIMAL GENERATORS, AND BORDERS

A pattern X is *closed* [160] in a dataset D if there is no proper super-pattern Y of X satisfying $\mathsf{count}(Y, D) = \mathsf{count}(X, D)$, or equivalently, $\mathsf{mds}(Y, D) = \mathsf{mds}(X, D)$. Closed patterns can be used to reduce the number of (frequent) patterns. For example, they can represent all frequent patterns in a lossless manner (allowing us to recover the supports of all frequent patterns from the closed patterns and their supports).

The *equivalence class* of a pattern X for a dataset D is defined as the set of all patterns Y satisfying $\mathsf{mds}(Y, D) = \mathsf{mds}(X, D)$. Such equivalence classes are *convex*, meaning that Z is in an equivalence class if there exist U and V in the equivalence class satisfying $U \subseteq Z \subseteq V$.

An equivalence class has exactly one maximal pattern, which is referred to as the *closed pattern* of the equivalence class, and a set of minimal patterns, which are referred to as the *minimal generators* of the equivalence class.

More generally, a convex set S of patterns can be represented by a *border* [56] of the form $< L, R >$, where L is the set of the minimal patterns (minimal in the set-containment relationship) of S and R is the set of maximal patterns of S. Clearly L and R are both anti-chains with respect to the set containment relation, in other words, there are no distinct patterns X and Y of L satisfying $X \subseteq Y$ and similarly for R. So an equivalence class is a special case of a border.

A border $< L, R >$ represents the collection $\{Z \mid X \subseteq Z \subseteq Y, X \in L, Y \in R\}$ of patterns. The patterns in such a collection do not need to have the same mds. This makes borders more flexible than equivalence classes.

The concept of closed patterns was introduced in [160]. The concepts for equivalence classes were presented in [24, 223]. Reference [115] argued that using minimal generators to represent equivalence classes is better than using closed patterns. Several methods in later chapters make use of minimal generator patterns. Reference [55] discussed an approach to represent a large number of minimal generators concisely.

2.6 ILLUSTRATING EXAMPLES

Table 2.1a gives an example of a relational dataset, and Table 2.1b gives binnings of the numerical attributes (using two bins for each attribute). For $g1$ the equi-width binning and entropy-based binning produce the same result. Table 2.2c gives the discretized version, and Table 2.2d the itemized version, of the relational data of Table 2.1a. In the discretized data for all attributes, the first interval is represented by 0 and the second by 1. To get the itemized data, the two intervals of $g1$ are represented as 0 and 1, respectively, the two intervals of $g2$ are represented as 2 and 3, respectively, the two intervals of $g3$ are represented as 4 and 5, respectively, and the two classes are represented as 6 and 7, respectively.

Table 2.1: Example of relational data and binning

TID	g1	g2	g3	Class
t1	-1	-99	50	POS
t2	0	100	20	POS
t3	1	999	80	NOR
t4	2	500	10	NOR

(a) Relational Data

GID	Bins	Encodings
g1	$(-\infty, 0.5), [0.5, +\infty)$	0, 1
g2	$(-\infty, 549), [549, +\infty)$	0, 1
g3	$(-\infty, 45), [45, +\infty)$	0, 1

(b) Binnings of Attributes

Table 2.2: Example of discretized and itemized data

TID	g1	g2	g3	Class
t1	0	0	1	POS
t2	0	0	0	POS
t3	1	1	1	NOR
t4	1	0	0	NOR

(c) Discretized Data

TID	g1	g2	g3	Class
t1	0	2	5	6
t2	0	2	4	6
t3	1	3	5	7
t4	1	2	4	7

(d) Itemized Data

For the itemized data in Table 2.2d with *minSup* = 2, the frequent patterns are: 0, 1, 2, 4, 5, 6, 7, 02, 06, 17, 24, 26, 026; the pattern 2 has support of 3 and all others have the support of 2. These frequent patterns belong to the following 5 equivalence classes: $< \{0, 6\}, 026 >$, $< \{1, 7\}, 17 >$, $< \{4\}, 24 >$, $< \{2\}, 2 >$, $< \{5\}, 5 >$. In the equivalence class of $< \{0, 6\}, 026 >$, 026 is the closed pattern, and 0 and 6 are its minimal generators; this equivalence class contains these patterns: 0, 6, 02, 06, 026; the matching dataset of every pattern in this equivalence class is $\{t1, t2\}$. The pattern 02 (on the discretized transactional data) corresponds to the pattern $g1 \in (-\infty, 0.5)$ & $g2 \in (-\infty, 549)$, which can be simplified to $g1 < 0.5$ & $g2 < 549$.

CHAPTER 3

Emerging Patterns and a Flexible Mining Algorithm

This chapter starts with the basics for emerging patterns. It then presents a simplified version of an emerging pattern-mining algorithm, which is easy to implement and can even be applied by hand on small datasets. Then it discusses what emerging patterns can represent in the context of group difference analysis. It also discusses the relationship between emerging patterns and association/classification rules, and the relationship among measures on patterns/rules used here and elsewhere. It then points to parts of the book that illustrate various uses of emerging patterns. This is followed by a discussion on traditional ways to analyze group differences. It concludes with a discussion of related issues.

In the literature, emerging patterns are also called *contrast patterns*.

3.1 SETTING FOR GROUP DIFFERENCE ANALYSIS

Throughout the book, by a *data group* we mean a dataset, namely a set of data instances. The dataset can be explicitly given by the user or implicitly defined using some conditions. Often a data group is linked to some property/aspect of the data of particular interest to the user. In general, a data group can be a class, or a (potential) cluster of data instances. It can also be a subset of data instances where a given regression/classification model makes large prediction errors. For concrete examples, a data group can be the positive class of colon cancer, a set of toxic molecules, a set of instances describing where and when some street crimes happened, a set of musical pieces of a particular type, and so on. Later chapters, especially Chapter 11, give many more examples of data groups.

For group difference analysis, an analyst may want to analyze two given data groups, focusing on the differences between the two groups. We will use the term *group difference* to refer to any expression or statement describing the differences between the two groups. Ideally, the expressions and statements representing group differences should be easy to understand and succinct. In this book we focus on group differences described by patterns.

3.2 BASICS OF EMERGING PATTERNS

Intuitively, emerging patterns are patterns whose supports in two given data groups differ significantly. Due to their support difference, such patterns can be used to discriminate the two data groups, and they can give indications on the ways the two data groups are different and the ways one group is unique. Many other uses can be derived from these capabilities.

There are at least three ways to formalize "supports differ significantly": a growth-rate based, a support-delta based, and a two-supports based. Growth rate has also been referred to as support ratio. Each of the three ways has some advantages. The growth rate based definition is intuitive, and as will be seen later, it is also flexible, allowing multiple uses of the corresponding patterns not allowed by the other ways.

Let D_1 and D_2 be two data groups and i, j be two integers in $\{1, 2\}$ such that $i \neq j$. The *growth rate*[1] [56, 57] of a pattern X from D_i to D_j is

$$\text{growthRate}(X, D_j) = \frac{\text{supp}(X, D_j)}{\text{supp}(X, D_i)}.$$

For the case $\text{supp}(X, D_i) = 0$, it is customary to

- define $\text{growthRate}(X, D_j)$ to be 0 if $\text{supp}(X, D_j) = \text{supp}(X, D_i) = 0$ and

- define $\text{growthRate}(X, D_j)$ to be ∞ if $\text{supp}(X, D_j) > 0$ and $\text{supp}(X, D_i) = 0$.

Clearly growthRate can be used to reflect the discriminativeness of patterns; patterns with larger growthRate values are more discriminative than patterns with smaller growthRate. In fact, as $\text{supp}(X, D_i)$ is the probability of X in group D_i, growthRate measures the change of group-based probabilities of X.

We now use growthRate to define emerging patterns.

Definition 3.1   Given a growth-rate threshold $\rho > 0$, a pattern X is a ρ-*emerging pattern* for data group D_j if $\text{growthRate}(X, D_j) \geq \rho$. If X is an emerging pattern for D_j, then D_j is the *home data group* (also called *target data group*), and the other dataset is the *opposing data group* (also called *background data group*) of X. An emerging pattern having $\text{growthRate}(X, D_j) = \infty$ is called a *jumping emerging pattern* for D_j.

In general, the largest growthRate of non-jumping emerging patterns of D_j is $\frac{1}{1/|D_i|} = |D_i|$; the largest growthRate of non-jumping emerging patterns (of D_1 or D_2) is $max\{|D_1|, |D_2|\}$. When D_1 and $D2$ have equal sizes n, the largest growthRate of non-jumping emerging patterns is n.

We will often write "(jumping) emerging patterns for D_2 vs. D_1" to refer to the (jumping) emerging patterns whose home group is D_2. We will also use EP as shorthand for emerging pattern, and JEP for jumping emerging pattern.

Consider the two data groups given in Table 3.1 and $\rho = 2$. Then $\{g1 = 0\}$ is an emerging pattern (whose growthRate is 2), and $\{g1 = 0, g2 = 2\}$ is a jumping emerging pattern, both with group 1 being the home group. But $\{g3 = 4\}$ is not an emerging pattern (whose growthRate = 1). Many other emerging patterns exist.

---

[1]One may use Laplacian smoothing to change ∞ to very large numbers while still having higher growthRate values for jumping emerging patterns than for all other emerging patterns.

Table 3.1: Two small data groups

TID	g1	g2	g3	Group ID
t1	0	2	5	1
t2	0	2	4	1
t3	1	3	4	1
t4	0	3	4	2
t5	1	2	4	2
t6	1	3	5	2

As emerging patterns can show how two data groups differ from each other and how a data group is unique when compared to another group, they are highly useful to solve a variety of problems, as will be seen in other chapters of this book.

Two approaches have often been used to select subsets of highly useful emerging patterns. One is to use the equivalence class approach, selecting a minimal generator for each equivalence class. The other is to select all minimal (jumping) emerging patterns (minimal in the set containment sense). In fact, Reference [112] stated that "the minimal jumping emerging patterns are the most expressive distinguishing patterns for classification;" they have also been referred to as the most general distinguishing patterns. The most general part comes from the fact that the patterns are minimal in the set containment sense and no proper subsets of those patterns are also jumping emerging patterns. In general, EPs with large growthRate improvement over all proper sub-patterns are also often useful; these patterns give a large "lift of power" over their sub-patterns. The last statement applies to jumping emerging patterns as well as other emerging patterns.

The *support delta* [25, 27] of a pattern X for data group D_j is defined as $\mathsf{supp}_\delta(X, D_j) = \mathsf{supp}(X, D_j) - \mathsf{supp}(X, D_i)$, where $i \in \{1, 2\} - \{j\}$. One can use this support delta to define support-δ-based patterns to express group differences. The *two-support-based method* uses two support thresholds, one being a support threshold $\alpha \in [0, 1]$ for the home data group and the other being a support threshold $\beta \in [0, 1]$ for the opposing data group. A pattern X is a (α, β)-*emerging pattern* for data group D_j if $\mathsf{supp}(X, D_j) \geq \alpha$ and $\mathsf{supp}(X, D_i) \leq \beta$ ($i \in \{1, 2\} - \{j\}$). Reference [56] used a number of (α, β) pairs to divide the "support plane" of the growthRate-based emerging patterns and to approximate the whole set of ρ-emerging patterns.

It is easy to define and visualize jumping emerging patterns when the growth-rate approach is used: there is no need to provide support thresholds, and one just needs to know that these patterns match some instances of one data group but do not match any instance of the other data group. The growth-rate approach to define jumping emerging patterns also offers some other advantages; for example, jumping emerging patterns without minimum support threshold are useful for outlier detection and intrusion detection (see Chapter 6). When the two-support-based approach is used, one only can define jumping emerging patterns that are frequent (in the home data group). When the support-delta-based approach is used, it is not possible to define jumping emerging patterns.

As will be seen later, support, growthRate, and length of emerging patterns all play a role on their usefulness. Everything else being equal, higher growthRate leads to more discriminative patterns; higher support leads to emerging patterns useful for characterization, classification, and clustering; low-support emerging patterns can be useful for the detection of novelty, outlier, and emerging trends; and short emerging patterns are preferred as they are easy to detect and can be indications of high quality of classes and clusters. In general, patterns of length 1 typically have low growthRate and hence typically are not very useful.

The growthRate measure is quite intuitive for measuring the discriminativeness of patterns. It was used in quite a few papers. For example, reference [2] used the growthRate measure to indicate the discriminant score of patterns in the design of a system for visual analysis of river water quality. Reference [206] also used the growthRate measure to select areas having unusual crime rates and to visualize such areas on maps. Reference [36] used growthRate to evaluate deviant gaming behavior and to analyze where a player erred in a MOBA game.

Remark: it also makes sense to refer to emerging patterns as *group distinguishing patterns* or *group distinguishing rules*. As mentioned earlier, emerging patterns have also been frequently called "contrast patterns" in the literature, especially since 2007.

3.3   BORDERDIFF: A SIMPLE, FLEXIBLE EMERGING PATTERN MINING ALGORITHM

References [56, 57] introduced a method called MBD-LLBORDER for mining emerging patterns. The algorithm itself relies on repeated use of a so-called BorderDiff function on two data groups, with one group consisting of just one instance. This function finds jumping emerging patterns that occur in the one instance but not in any instances in the other data group.

This section presents a simplified version of that BorderDiff function, which is denoted by BorderDiffSimp here; BorderDiffSimp omits various optimization techniques used in BorderDiff. One particular strength of the BorderDiffSimp algorithm is that it is easy to use by hand on small datasets.

We now describe the BorderDiffSimp algorithm. Given an instance t and an opposing set S of instances, BorderDiffSimp first computes the set difference between t and each s in S. Then it performs a sequence of expansion-and-minimization steps as follows: it makes a set of singleton patterns from the items in (the first) one of the set differences. Then it iteratively expands the set of current minimal patterns by joining that set with the next set difference d_i; this result of the join is a set that consists of all possible unions of one of the current minimal patterns and a singleton set constructed from one item in d_i. The minimization step is then performed, removing non-minimal patterns from the result of the join. The above is repeated until all set differences have been incorporated in the expansions. The pseudo-code of the algorithm is given in Algorithm 3.1.

Algorithm 3.1 BorderDiffSimp Algorithm

Input: t (a target instance); S (a set of opposing instances)
Output: the minimal jumping emerging patterns for t vs. S
  (1) For each $s_i \in S$, let $d_i = t - s_i$
  (2) Let $MEP = \{\{x\} \mid x \in d_1\}$
  (3) For each i ($2 \le i \le |S|$) do
    (4) Expansion: Let $MEP = \{X \cup \{x\} \mid X \in MEP, x \in d_i\}$
    (5) Minimization: Remove non-minimal patterns of MEP
  (6) End (For)
  (7) Return MEP

To illustrate, consider $t = 1234(= \{1, 2, 3, 4\})$ and $S = \{3456, 2479, 2358\}$. Then step (1) produces

$$d_1 = 1234 - 3456 = 12, \quad d_2 = 1234 - 2479 = 13, \quad d_3 = 1234 - 2358 = 14.$$

In step (2), from $d_1 = \{1, 2\}$, MEP is initialized to $\{\{1\}, \{2\}\}$. For d_2, the expansion operation given by step (4), joining MEP with $d_2 = \{1, 3\}$, produces $MEP = \{\{1\}, \{1, 2\}, \{1, 3\}, \{2, 3\}\}$. (Observe that $\{1, 1\} = \{1\}$.) Removing non-minimal patterns using step (5), MEP becomes $\{\{1\}, \{2, 3\}\}$. For d_3, the expansion operation given by step (4), joining MEP with $d_3 = \{1, 4\}$, produces $MEP = \{\{1\}, \{1, 2, 3\}, \{1, 4\}, \{2, 3, 4\}\}$. Removing non-minimal patterns using step (5), MEP becomes $\{\{1\}, \{2, 3, 4\}\}$. The algorithm finally outputs two minimal JEPs: 1 and 234.

If we do not do the iterative expansion-minimization steps, then one may need to generate and minimize the following set of patterns: $\{\{x_1, x_2, \ldots, x_{|S|}\} \mid x_1 \in d_1, \ldots, x_{|S|} \in d_{|S|}\}$. Each pattern in this set contains an item in t but not in s_i for each i; hence the minimal patterns in this set are precisely the set of JEPs that occur in t but not in any s_i. When $|S|$ is in the hundreds and t is large (e.g., > 30), the set becomes too large to work with. The iterative expansion-minimization steps help reduce the impact of the combinatorial explosion and still find the same set of minimal JEPs.

The minimization steps are often very expensive and many techniques can be introduced to speed up the computation. Of course many newer algorithms are faster than BorderDiff, but they are not as simple as BorderDiff, cannot be easily used in pencil-paper computations, and are not as flexible as BorderDiff (see below).

One major strength of the BorderDiff algorithm is that it is flexible—it can be used on dynamically selected subsets of data with dynamically selected sets of attributes. This makes it very useful when the number of instances or the number of attributes is too large and one wishes to mine highly discriminative emerging patterns using dynamically selected samples of instances or dynamically selected subsets of attributes. We illustrate this with two ways to use the algorithm.

- The BorderDiff algorithm is useful for intrusion detection and outlier detection methods that evaluate the unusualness of an instance based on the minimal JEPs contained in the instance but the minimal JEPs are not contained in a random sample of normal instances (see Chapter 6). Here, one of the two groups exactly contains a testing instance which one wishes to check if it is an outlier, and the other group is a randomly selected subset of the normal instances. The idea of using a random subset was adopted to increase the robustness as well as efficiency of the outlier detection method (see Chapter 6). It is hard for other algorithms to mine jumping emerging patterns in this situation.

- The BorderDiff algorithm is useful for mining high-quality emerging patterns (and also ranking genes for complex diseases) in the presence of thousands of attributes/genes by dynamically applying BorderDiff on a small dynamically selected set of attributes for each of some k top-ranked attributes in an iterative manner (see Chapter 9). Since the number of attributes is too large, no known algorithm can effectively mine emerging patterns by considering all attributes at once. The BorderDiff algorithm was used since it can be flexibly used on dynamically selected subsets of attributes.

It should be noted that one can mine emerging patterns that are not jumping emerging patterns by repeatedly using BorderDiff on multiple pairs of "one random instances t (in one group) vs. a random sample of instances S in another group."

Incidently, we note that it is often infeasible to mine jumping emerging patterns by first mining frequent patterns. For example, suppose the data instances, given by t (one instance) and S (a set of instances), have reasonably high dimensionality (e.g., between 50 and 100) and the set S has a reasonably large number of instances (e.g., 500). Observe that all subsets of t have 100% support for the group of $\{t\}$. To mine all jumping emerging patterns by working with frequent patterns, we need to know all frequent patterns of S for $minSup = 1/500$. This threshold is too small to work with for frequent pattern mining algorithms.

3.4   WHAT EMERGING PATTERNS CAN REPRESENT

Generally speaking, emerging patterns represent group differences contrasting given data groups/classes. Such representation enables analysts to not only analyze group differences, but also make use of group differences to solve various problems.

- Emerging patterns can represent *distinguishing patterns/conditions* for data groups/classes: emerging patterns match (many) more instances in their home groups/classes than in other groups/classes, so they can be used to distinguish instances of their home groups/classes against those from other groups/classes. This aspect is the basis of the following two items.

- Emerging patterns can represent *unique properties, characterizations, and signatures* of their groups/classes: jumping emerging patterns, and emerging patterns with extremely high growth rates, have the above three capabilities since they do not match any (or many) instances of the opposing groups. They characterize their home groups/classes and can serve as signatures of their home groups/classes. A large number of papers that make use of emerging patterns in this fashion are briefly discussed in Chapter 11.

- Emerging patterns can represent *multi-factor interactions* linked to classes/states associated with the underlying problem represented by the data. Due to their high discriminative power and because they represent unique properties of their home classes, jumping emerging patterns, and emerging patterns with extremely high growth rates, can represent complex interactions of complex diseases/problems. The fact that the component conditions inside such patterns must all be present to give the patterns high discriminative power (in other words, all proper subsets have significantly lower discriminative power) makes these patterns the best choices representing complex interactions. Observe that we generalize the concept of complex interactions to other complex problems (typically in non-medical disciplines) that are similar to complex diseases from a pattern-theoretic perspective. More discussion on complex interactions can be found Chapter 9. We believe that many practical problems are complex and emerging patterns' usefulness can perhaps be attributed to the fact that they represent complex interactions and that they allow us to make use of discriminative complex interactions.

3.5 COMPARISON WITH ASSOCIATION RULES, CONFIDENCE, AND ODDS RATIO

Emerging patterns are related to association rules and classification rules, and the growthRate measure is related to other measures that are used to indicate "strength" of rules and patterns. This section gives a brief discussion on these relationships.

Clearly, emerging patterns are related to association/classification rules. Indeed, "P is a potential emerging pattern for a group $G2$" is equivalent to "$P \rightarrow G2$ is a potential association/classification rule." The word "potential" is used above since thresholds are needed to define emerging patterns and association rules.

We now discuss how the growthRate measure is related to other measures, namely confidence of association rules and Odds Ratio (of patterns with respect to a class).

We consider the case where there are two groups, $G1$ and $G2$. So the underlying dataset D is the union of the two groups. All three measures are defined based on some basic counts (given in Table 3.2) and also some derived counts.

Assuming that $G2$ is the group of interest for emerging patterns and association rules, then the three measures are defined as follows.

Table 3.2: Contingency table

	$G1$	$G1$
P	$count(P,G1)$	$count(P,G2)$
$\neg P$	$count(\neg P,G1)$	$count(\neg P,G2)$

- Confidence [5] for the association rule $P \rightarrow G2$ is defined by

$$Conf(P \rightarrow G2) = \frac{count(P,G2)}{count(P,G1 \cup G2)} = \frac{count(P,G2)}{count(P,G1) + count(P,G2)}.$$

- growthRate of P (as EP of $G2$) is defined by

$$\text{growthRate}(P,G2) = \frac{\text{supp}(P,G2)}{\text{supp}(P,G1)} = \frac{count(P,G2)/(count(P,G2) + count(\neg P,G2))}{count(P,G1)/(count(P,G1) + count(\neg P,G1))}.$$

- *OddsRatio* of P (with respect to $G2$) is defined[2] as

$$OddsRatio(P) = \frac{count(P,G2)/count(\neg P,G2)}{count(P,G1)/count(\neg P,G1)}.$$

We note that growthRate(P,G2) measures the change of group-based probabilities of P, as supp(P, Gi) is the probability of P in group Gi. Moreover, *OddsRatio(P)* measures the change of the odds of P in the groups.

Among the three measures we see the following three relationships.

(a) Clearly growthRate and *OddsRatio* are very similar. Their difference lies with the following: growthRate is the fold of change in probabilities (or matching data portion size) in the two groups, whereas *OddsRatio* is the fold of change in the odds (of matching the pattern) in the two groups. Based on this difference, it can be argued that growthRate is more intuitive than *OddsRatio*, as growthRate is related to a more primitive concept (probability or portion), whereas *OddsRatio* is related to a less primitive concept (odds). So growthRate and *OddsRatio* both measure the change of probability of P, but in slightly different ways.

(b) Moreover, growthRate of P (as EP of $G2$) is positively correlated with the confidence of $P \rightarrow G2$.

---

[2]Odds Ratio is defined as the ratio of the odds of $G2$ (an outcome) in the presence of P (an exposure) and the odds of $G2$ without the presence of P. This statistic attempts to quantify the strength of the association between $G2$ and P. If the ratio is greater than 1, then $G2$ is considered to be associated with P in the sense that, compared to the absence of P, the presence of P raises the odds of $G2$. See https://en.wikipedia.org/wiki/Odds_ratio.

(c) On the other hand, confidence of association rules may not be very informative when the two groups have imbalanced sizes. Indeed, if $G2$ is the minority class in imbalanced classification, then confidence of $P \rightarrow G2$ can be very small even when $\text{supp}(P, G2)$ is large. The confidence measure, as a function of $count(P, G2)$ and $count(P, G1 \cup G2)$, can be more strongly influenced by $count(P, G1)$ (in the majority class) than by $count(P, G2)$ (in the minority class).

Table 3.3: Comparison: growthRate, Odds Ratio, confidence for P and $G2$

$|G1|$	$|G2|$	$count(P,G1)$	$count(P,G2)$	GR	OR	Conf
100	1,000	20	50	0.25	0.21	0.71
1,000	1,000	20	50	2.5	2.58	0.71
1,000	100	20	50	25	49.00	0.71
100	1,000	20	100	0.5	0.44	0.83
1,000	1,000	20	100	5	5.44	0.83
1,000	100	20	100	50	∞	0.83
100	1,000	50	50	0.1	0.05	0.50
1,000	1,000	50	50	1	1.00	0.50
1,000	100	50	50	10	19.00	0.50

Table 3.3 illustrates the relationships among the three measures. It contains nine rows with different group-size ratios and different matching data-size ratios. We can see that (i) the confidence measure is constant for a given matching dataset size setting (the third and fourth columns) when the underlying group sizes (the first and second columns) vary, and hence it does not properly reflect the portion size (probability) difference of P in the two groups, and (ii) both growthRate and *OddsRatio* do not suffer from the above weakness.

3.6   POINTERS TO SECTIONS ILLUSTRATING USES OF EMERGING PATTERNS

Due to the capabilities discussed above, emerging patterns have been used in many ways. The following are some references to relevant parts of the book illustrating various uses; clearly these examples are not exhaustive.

- Emerging patterns have been used as features (e.g., Section 11.14), regression terms (Section 11.8), basic classifiers (Chapter 4), toxicity alert triggers (Section 11.4), disease biomarkers (Section 11.3), disease subtype definitions (Section 11.5), and subpopulation characterizations (Chapter 10).

- Emerging patterns have been used for class/group analysis and understanding (Sections 11.1 to 11.7), classification (Chapter 4, Sections 11.15–11.17, Section 11.19, Section 11.18), clustering (Chapter 7) and clustering quality evaluation (Chapter 8), outlier detection and novelty discovery (Chapter 6), gene ranking for complex diseases (Chapter 9), classification/regression model analysis and improvement (Chapter 10), and compound screening and drug candidate selection (Chapter 5).

- Emerging patterns have been used in diverse applications including crime analysis (Section 11.6), activity recognition (Section 11.13, Section 11.16, Section 11.10), music analysis (Section 11.7), medicine/bioinformatics (Section 11.3, Section 11.5, Section 11.8, Section 11.9, Section 11.11, Section 11.15), chemoinformatics (Section 11.1, Section 11.2, Section 11.4), and so on.

3.7   TRADITIONAL ANALYSIS OF GROUP DIFFERENCES

Most traditional approaches to analyzing group differences rely on the use of group prototypes or group profiles; users need to look at the respective profiles or prototypes to appreciate the difference between the groups. One example is to represent each data group using the mean instance of the group. When the data groups have high dimensions, these mean instances are long vectors, making it hard to identify useful discriminative information. Another example is to represent each data group using some graphically displayed charts such as histograms, boxplots, and bar charts, where different attributes are often associated with separate charts; it is difficult to spot useful patterns concerning how the groups differ, or to use them to form solutions for data analysis problems.

Several statistical tests have been developed to determine if two groups are different (see, e.g., [144]), including one based on checking if two given data groups have different group means. Clearly, performing such statistical tests will not give any concrete insightful pattern[3] showing how the groups differ.

Other ways to compare two data groups include comparing models developed from the two data groups using an identical algorithm. The model can be regression based, classification based, or clustering based. A related approach is to evaluate the difference between two data groups by using frequent patterns derived from one data group to compress the other data group [204].

Reference [175] recognized the need to go "beyond differences in means" and introduced some graphical methods to compare two groups. This work stated that going "beyond differences in means" is one of some changes that are necessary to improve the quality of neuroscience research.

---

[3]We note that [123] helped to fill the gap by considering exploratory hypothesis testing and analysis through the discovery of patterns.

In bioinformatics there are many studies involving the discovery of DNA or protein sequence differences after aligning sequences of different sequence families (e.g., [121]).

It appears that traditional approaches to analyzing group differences do not provide information that concisely indicates how the given data groups differ. In contrast, emerging patterns give concise descriptions of how the given data groups differ; importantly, these patterns can give insights on the difference, and they are "usable" for formulating solutions for a vast array of data analysis problems.

3.8   DISCUSSION OF RELATED ISSUES

While many emerging/discriminative pattern mining algorithms have been developed, several such algorithms are useful in some particular ways. (a) Reference [116] gave an algorithm for mining equivalence classes of emerging patterns, which has been used by quite a few recent papers. (b) The emerging pattern mining algorithm of [84] is somehow special as it does not make use of *a priori* numerical attribute binnings; instead, it forms such binnings in the mining process and mines the EPs using a diverse set of decision trees that involve some randomness in the tree-building process. (c) References [44, 179] gave algorithms for efficiently mining a concise set of discriminative patterns for use in classification model building or generalized linear model building. We also note that the BorderDiff algorithm is unique in that it can be flexibly used in dynamic settings, is useful for mining jumping EPs as well as general EPs, and it can used by hand on small datasets.

Researchers have also proposed and used several special kinds of emerging patterns, including ones that are stable [141], strong [185], essential [70], and natural [6]. Several chapters in [51] discuss quality measures of (emerging) patterns and survey emerging pattern mining algorithms.

Chapter 11 includes quite a few data analysis applications based on using emerging patterns to capture group differences. Earlier in the writing process we planned to include these applications in this chapter, but at the end we decided to put them in Chapter 11 instead in order to keep this chapter's length under control.

Mining emerging patterns has very high theoretical computational complexity. Indeed, it was shown [208] that finding a jumping emerging pattern P for two given data groups D_1 and D_2 maximizing $\mathsf{supp}(P, D_2)$ is MAX SNP-hard, which implies that polynomial time approximation schemes do not exist for the problem unless $P = NP$. However, experiments demonstrate that the BorderDiff algorithm can often finish quickly on real world data (see [57] and Chapters 6 and 9).

It is interesting to note that the BorderDiff algorithm can be used to compute minimal transversals of hypergraphs [22, 198].

As an aside, we note that [58] introduced the concept of "conditional contrasts" to capture situations where a small change in patterns is associated with a big change in the matching data of the patterns. A conditional contrast is a triple (B, F_1, F_2) of three patterns, with B being its condition/context pattern, and F_1 and F_2 being its contrasting factors. Such a conditional contrast is of interest if the syntactic difference between F_1 and F_2 is relatively small, and the difference between the corresponding matching datasets $\mathsf{mds}(B \cup F_1)$ and $\mathsf{mds}(B \cup F_2)$ is relatively large. This direction is not related to group difference in the sense discussed above.

CHAPTER 4

CAEP: Classification By Aggregating Multiple Matching Emerging Patterns

Many emerging patterns can be mined from the training data of given classification problems. Each such pattern captures some class-discriminating signals that can be used when building pattern-based classification models. The challenge is that one needs to decide how to select and use such patterns, and in doing so one must consider the discriminativeness and coverage of individual patterns as well as similarity and matching dataset overlap among patterns. The choices made will significantly impact on the quality of the constructed pattern-based classification model.

This chapter focuses on a discriminating pattern aggregation based approach, called CAEP, to pattern-based classification. CAEP pioneered the idea that "the class-discriminating signals contained in multiple patterns matching a given instance should be combined in order to build an accurate and robust classifier." CAEP belongs to the family of rule/pattern-based classifiers. CAEP has been used by many researchers in various applications from many disciplines, and many variants of CAEP have been proposed.

Organizationally, this chapter first gives some background materials on classification. Then it presents the CAEP approach, followed by a discussion on CAEP's performance in experiments and in applications, as well as its uniqueness and strengths. Then it presents a lazy instance-based approach, called DeEPs, which also follows the aggregation approach. It then discusses CAEP's relationship with other rule/pattern-based classifiers. Finally, some concluding remarks are given.

This chapter is related to two other chapters. Chapter 5 discusses CAEP's capability for classification when the training datasets are tiny, and the exploitation of that capability for effective candidate compound selection for drug design and so on. Chapter 10 presents another way of using EPs for regression and classification. While this chapter uses emerging patterns (EPs) as basic conditions/classifiers, Chapter 10 uses EPs as subpopulation characterization to identify exploitable boosting opportunities and to build pattern aided regression and classification models.

4.1   BACKGROUND MATERIALS ON CLASSIFICATION

Classification is a supervised learning task. Its training dataset is a finite set of some s pairs of instance and class of the form $\{(x_i, y_i) \mid 1 \leq i \leq s\}$. The aim is to learn a classification model to predict y from x. A classification model is often called a classifier.

A very large number of classification algorithms, and many performance evaluation measures, have been proposed. More details can be found in textbooks and survey papers.

Main challenges for classification include high dimensionality of data, imbalanced class distribution, behavior heterogeneity of data instances (called diverse predictor-response relationships in Chapter 10), tiny training data, and small disjuncts.

Imbalanced classification is concerned with classification where the underlying data have an imbalanced class distribution. There are many more data instances in certain majority classes than in the minority (rare) classes. As noted in [190], the class imbalance problem is pervasive in a large number of domains of great importance in data mining. Most traditional classification algorithms assume a relatively balanced class distribution. Imbalanced classification is a major challenge to them, as they often cannot produce classifiers having acceptable accuracy. Techniques to handle imbalanced classification include data-level approaches (balancing the training data by sampling, etc.), algorithm-level approaches, and cost-based approaches [190].

It is a challenge for classification algorithms to effectively tackle the "small disjuncts" [95], each of which characterizes a small region in the data space containing very few data points, and the class labels of those data points are mostly different from the class labels of the majority of their neighboring points. Reference [214] showed that, for some 30 real-life datasets, classification errors are often heavily concentrated toward the smaller disjuncts.

The tiny training data challenge is concerned with classification problems where the training dataset is tiny (e.g., 3 instances in class 1 vs. 3 instances in class 2, for a two-class problem). It turns out this is a challenge for traditional methods. See Chapter 5 for more details.

4.2   THE CAEP APPROACH

A CAEP classifier uses a set $\mathcal{E}$ of selected emerging patterns (EPs). Let $C_1, \ldots, C_\kappa$ be the classes. Let $\mathcal{E}_i$ denote the subset of EPs in $\mathcal{E}$ whose home class is class C_i. The CAEP algorithm's performance is affected by the set of emerging patterns mined and selected in the training phase.

Two prominent features of the CAEP classification algorithm are the following:

(1) Aggregating/combining the discriminative power of all matching emerging patterns (from a set selected for use by the classifier) of a given instance to compute the class-likelihood scores for the instance.

(2) Using a normalizing technique to adjust the class-likelihood scores in order to address class imbalance and pattern imbalance issues.

Sections 4.2.1 and 4.2.2 below discuss how to compute and normalize the likelihood scores respectively. Section 4.2.3 addresses the pattern set selection issue. Section 4.2.4 presents the CAEP training and testing algorithms.

4.2.1   CAEP'S CLASS-LIKELIHOOD COMPUTATION

Given an instance t to classify, CAEP computes a likelihood score for "t belongs to C_i" for each class C_i. This is done by aggregating the class-discriminating signals contained in all patterns in $\mathcal{E}_i$ that match t. CAEP measures the class-discriminating signal of a pattern P using $\mathsf{supp}(P, C_i) * \frac{\mathsf{growthRate}(P,C_i)}{\mathsf{growthRate}(P,C_i)+1}$. This value is proportional to both $\mathsf{supp}(P, C_i)$ and $\frac{\mathsf{growthRate}(P,C_i)}{\mathsf{growthRate}(P,C_i)+1}$. This value is the vote by P for "t belongs to C_i." So CAEP lets EPs with larger $\mathsf{supp}(P, C_i)$ and larger $\frac{\mathsf{growthRate}(P,C_i)}{\mathsf{growthRate}(P,C_i)+1}$ to give a larger vote.

CAEP computes the (raw) aggregated likelihood for t belonging to C_i by

$$\mathsf{CAEPLSr}(t, C_i) = \sum_{P \in \mathcal{E}_i, P \; matches \; t} \mathsf{supp}(P, C_i) * \frac{\mathsf{growthRate}(P, C_i)}{\mathsf{growthRate}(P, C_i) + 1}.$$

To understand the $\frac{\mathsf{growthRate}(P,C_i)}{\mathsf{growthRate}(P,C_i)+1}$ part, we use some $\mathsf{growthRate}(P, C_i)$ and $\frac{\mathsf{growthRate}(P,C_i)}{\mathsf{growthRate}(P,C_i)+1}$ value pairs to illustrate:

$$(1, 0.5), (2, 0.67), (4, 0.80), (6, 0.86), (8, 0.89), (12, 0.92), (24, 0.95), (48, 0.98), (\infty, 1).$$

So, for $\mathsf{growthRate}(P, C_i) \geq 1$, the value of $\frac{\mathsf{growthRate}(P,C_i)}{\mathsf{growthRate}(P,C_i)+1}$ is proportional to $\mathsf{growthRate}(P, C_i)$ and its range is between 0.5 and 1; the largest value is achieved when P is a jumping emerging pattern.

We use triples of the form $(\mathsf{growthRate}(P, C_i), \mathsf{supp}(P, C_i), \mathsf{supp}(P, C_i) * \frac{\mathsf{growthRate}(P,C_i)}{\mathsf{growthRate}(P,C_i)+1})$ to get an intuitive feel of the votes given by patterns with different characteristics:

$$(4, 0.25, 0.2), (8, 0.2, 0.18), (24, 0.01, 0.1), (48, 0.05, 0.049), (\infty, 0.03, 0.03).$$

Observe that patterns of $\mathsf{growthRate} = 1$ have no discriminatig power for classification, so no triple with $\mathsf{growthRate} = 1$ is included in the list.

Importantly, even EPs with small support still give a fairly sizable vote, if they have large growthRate. There can be many EPs with large growthRate but small support, and their discriminating power is combined by CAEP. These EPs are often ignored by other clasification algorithms. Having a large EP set, CAEP often benefits from this use of low-support EPs. In particular, such EPs can capture "small disjuncts."

4.2.2   CAEP'S LIKELIHOOD NORMALIZATION

Sometimes the numbers of instances of different classes are highly imbalanced. For example, for many diseases, the number of tested patients who are positive is much smaller than the number of tested patients who are negative. Classification problems with this characteristic are commonly referred to as "imbalanced classification" problems.

There is another kind of imbalance, which is often overlooked in the literature: There are many more class-distinguishing patterns (or EPs) for one of the classes than for the other classes. Moreover, the patterns for that class may typically have much larger growthRate. We call this the "imbalanced pattern phenomenon;" it happens when some classes are "pattern rich" and the other classes are "pattern poor."

When using the raw aggregated scores to classify instances, an instance t of a pattern-poor class is often classified as an instance of a pattern-rich class. This is because the pattern-rich class has more patterns, which often have high discriminating power, matching t than does the pattern-poor class.

CAEP introduced a normalization technique to combat this problem as follows: For each class C_i, let D_i be its associated set of training data instances. Sort the set of instances in D_i based on their raw likelihood for C_i given by CAEPLSr. Let $baseLS(C_i)$ be the 85 (which can be changed to other values) percentile of the sorted values (so 85% of the values are less than or equal to $baseLS(C_i)$). CAEP uses the normalized score $CAEPLS(t, C_i) = CAEPLSr(t, C_i)/baseLS(C_i)$ to classify instances. That is, the class C_j such that $CAEPLS(t, C_j) = \max\{CAEPLS(t, C_i) \mid 1 \leq i \leq \kappa\}$ is the predicted class for t.

There are several other approaches to normalize the raw likelihood scores: (1) If numbers of EPs for the classes are small, then divide the raw likelihood scores for a class C by the sum of $sup * GrowthRate$ for all EPs associated with C. (This idea was used in [18].) (2) Normalize the sum of contributions of some top-k matching EPs of a class C by the sum of contributions of the top k EPs of class C. (This idea was used by PCL [117].)

4.2.3   EMERGING PATTERN SET SELECTION

The CAEP approach is based on the idea that all emerging patterns contain some class-discriminating signals, and it is designed to use such signals of as many emerging patterns as possible. However, including all possible emerging patterns is infeasible because often there are too many of them. Moreover, patterns that are very similar to each other (measured by their mds) will cast nearly identical votes for classifying all test instances. In the extreme case, two highly similar patterns can be viewed as one pattern giving two duplicated votes. To address this issue, a diverse set of patterns should be selected, and patterns of high quality should be preferred in general.

There are two important factors in pattern set selection: quality and diversity. High-quality patterns should be those that are used often and that cast high-confidence votes. Pattern quality is often evaluated in terms of growthRate and supp. Pattern set diversity is concerned with relationship among patterns, which can be reflected in mds overlap and quality difference among patterns with highly similar mds. Minimal EPs (in the set containment sense) are often given higher preference. EPs of high growthRate are often preferred over EPs of lower growthRate. A candidate EP P whose mds is very different from the mds of EPs that are already selected should receive high preference for being selected. If P's mds is very similar to the mds of some selected EP, it is common to select P only if its quality meets some thresholds on improvement (over those already selected that are very similar to P); growth rate improvement of EPs [231] can be used on this regard.

4.2.4   THE CAEP TRAINING AND TESTING ALGORITHMS

Algorithms 4.2 and 4.3 give the training and testing algorithms for CAEP. For the training part, the EPs for class C_i are mined from D_i vs. $\cup_{j \neq i} D_j$. In Algorithm 4.2, π, minSup, minGR, minSim, and minImp are thresholds for EP set selection concerning normalizing percentile, support, growthRate, similarity, and quality improvement, respectively. Details of the steps of these algorithms were discussed in the previous sections.

Algorithm 4.2 CAEP Training

Parameters: π (percentile for normalization)
   minSup, minGR, minSim, minImp (thresholds for EP set selection)
Input: $D_1, \ldots, D_\kappa$ (training data for κ classes)
Output: $\mathcal{E}$ (a set of EPs); $b_1, \ldots, b_\kappa$ (baseline values for normalizing CAEPLSr-scores)
   (1) Mine and select a diverse set $\mathcal{E}$ of high-quality EPs from $D_1, \ldots, D_\kappa$
   (2) Compute CAEPLSr-scores for each training instance and class combination
   (3) Find the π percentile of the CAEPLSr-scores $b_1, \ldots, b_\kappa$ for the classes
   (4) Return $\mathcal{E}$ and $b_1, \ldots, b_\kappa$

Algorithm 4.3 CAEP Testing

Parameters: $\mathcal{E}$ (a set of EPs);
   $b_1, \ldots, b_\kappa$ (baseline values for normalizing CAEPLSr-scores)
Input: t (testing instance)
Output: a class (predicted class for t)
   (1) For each class C_i, compute CAEPLS(t, C_i) using $\mathcal{E}$ and $b_1, \ldots, b_\kappa$
   (2) Return the class C_j maximizing CAEPLS(t, C_i) $(1 \leq i \leq \kappa)$

4.3   A SMALL ILLUSTRATING EXAMPLE

We now illustrate CAEP using the data given in Table 4.1. For simplicity, we consider using minimal JEPs only (in the set-containment sense). The set of all possible minimal JEPs are listed in Table 4.2. We take $\infty/(\infty + 1)$ as 1. So it is as if only the supports of JEPs are used in computing the likelihood scores.

Table 4.1: Small example illustrating CAEP

TID	A1	A2	A3	Class
t1	0	2	5	+
t2	0	2	4	+
t3	1	3	4	+
t4	0	3	5	-
t5	1	3	5	-
t6	1	2	5	-

Table 4.2: Jumping emerging patterns for data in Table 4.1

Pattern	$supp_+$	$supp_-$
02	2/3	0
04	1/3	0
14	1/3	0
24	1/3	0
34	1/3	0
03	0	1/3
12	0	1/3
15	0	1/3
35	0	1/3

Consider classifying the tuple $t = 124$ using the JEPs in Table 4.2. Since t matches the two JEPs 14 and 24 of the + class, t's score for the + class is $2/3 = 1/3 + 1/3$. Since it matches the JEP 12 of the − class, t's score for the − class is $1/3$.

To compute the normalized scores, we need to compute the raw scores for all instances in Table 4.1 to determine the desired percentile values.

Suppose we use all emerging patterns with growthRate > 1 (instead of just the JEPs). Then we also need to include several single item patterns, e.g., 2 whose growthRate $= 2$ and whose supp in the + class is $2/3$. For a tuple matching the pattern "2", the pattern contributes the score of $2/3 * 2/(2 + 1) = 4/9$.

4.4   EXPERIMENTS AND APPLICATIONS BY OTHER RESEARCHERS

References [63, 228] provided experimental performance evaluation of CAEP (using different types of selected EP sets). Reference [189] studied noise tolerance for EP-based classifiers. Reference [131] examined performance of contrast pattern based classifiers in imbalanced databases under different resampling strategies. Reference [38] studied the performance of extensions of CAEP in the multi-relational setting.

The CAEP approach was applied in many applications in non-computing disciplines; the following is a partial list. More details of some can be found in Chapters 5 and 11. Reference [18] used CAEP to work with tiny training datasets for effective candidate compound selection in exploratory drug design. Reference [158] used CAEP for the prediction and characterization of P-glycoprotein substrates potentially bound to different sites. (The above two references used the name of ECP (Emerging Chemical Pattern) for CAEP.) Reference [234] used CAEP to evaluate surgical performance and provide feedback (including identifying the stage of the surgery) during simulated ear (temporal bone) surgery in a 3D virtual environment. Reference [222] considered metamorphic malware detection based on aggregating emerging patterns. Reference [162] used CAEP for the diagnosis of myocardial ischemia.

4.5   STRENGTHS AND UNIQUENESS OF CAEP

4.5.1   STRENGTHS OF CAEP

- The typical pattern set used by CAEP is often not the result of a greedy search. In comparison, the decision tree approaches and non-aggregation rule-based classifiers often use a very small set of rules selected through a highly greedy search process. The aggregation based approach can use a highly diverse and large set of high-quality patterns, which often leads to accurate classification.

- Emerging patterns of low support are often used in CAEP, in addition to those of high support. This can help handle small disjuncts [95], namely regions of instances whose class distributions are highly different from the class distributions of their immediate surrounding regions.

- The aggregation approach of CAEP allows one to use all possible distinguishing signals in the form of patterns. This is a major advantage especially when useful distinguishing signals are rare. This is also confirmed by the fact CAEP has (relatively) good performance when the training dataset is tiny.

- The CAEP approach is also quite good for imbalanced classification problems, and it is also fairly noise tolerant.

4.5.2   UNIQUENESS OF CAEP

(1) Allowing the use of a diverse set of high-quality patterns: The pattern set can be large. This allows the pattern set to be a very diverse set; patterns with very small supports can be included. As each pattern represents an interaction among the attributes, the pattern set can include many distinctive interactions, and hence can cover many small disjuncts.

(2) The way to combine the votes by multiple matching patterns: The vote of a pattern in the CAEPLS value is influenced by two factors, namely the support of the pattern in its home class and the pattern's growth rate.

(3) CAEP's voting vs. traditional voting: CAEP's voting is a distinct type of voting. Indeed, each member of a traditional ensemble votes on all instances, whereas in CAEP a pattern votes only on matching instances. In CAEP, a pattern is not a standard classifier: it can have a very small support, hence it can only cast a vote on a very small fraction of the data instances. It does not cast a vote on a majority of instances. Moreover, on all of those instances where a pattern casts a vote, all of those votes are for the same class. In traditional ensemble voting, each ensemble member cast a vote on all instances, and different votes can be for different classes.

Because of the uniqueness discussed in (3), we use the term *simple classifier* to refer to the way a pattern is used in CAEP. (We also thought about using the term *one hot classifier* to denote this use. But we did not use this choice in the end since there is a major difference between one-hot and the way a pattern is used in CAEP, namely in one-hot two outcomes are represented as 0 and 1, and in CAEP, there is just one outcome or none.)

4.6   DEEPS: INSTANCE-BASED CLASSIFICATION USING EMERGING PATTERNS

The DeEPs classification algorithm [114] uses a lazy instance-based approach to classification. It is also an aggregation-based approach.

Let $C_1, \ldots, C_\kappa$ be the set of classes and let D_i be the dataset of class C_i in the training data. A discretization step is performed on the data if needed.

Given a test instance t, DeEPs first projects each D_i to $D_i^t = \{s \cap t \mid s \in D_i\}$, which amounts to removing from all $s \in D_i$ all items not in t. Then DeEPs mines the set of minimal jumping emerging patterns for each of classes, using their projected datasets, namely $D_1^t, \ldots, D_\kappa^t$. Let $\mathcal{E}_i$ denote that set for class C_i. The computed minimal jumping EPs are used to derive a so-called *compacted score* for each class C_i, defined by $compScore(t, Ci) = \frac{|mds_i|}{|D_i|}$, where $mds_i = \{s \in D_i \mid s \text{ matches some JEP in } \mathcal{E}_i\}$. If C_i is the class having the highest *compScore*, DeEPs then classifies t as a member of C_i.

DeEPs has several advantages. (1) By working with the projected training data relative to the test instance t, the data used in pattern mining has a significantly reduced volume and dimensionality. (Due to the nature of emerging pattern mining, only maximal tuples in the projected D_i^t are needed in the mining of emerging patterns.) (2) By mining emerging patterns with the projected datasets for given test instance t, one can find the most discriminative patterns contained in t, which may be lost in the pattern mining/selection process in other classification approaches, including the general CAEP approach. (3) Using the volume of matching datasets of JEPs of each class, the duplicated vote issue is avoided.

4.7   RELATIONSHIP WITH OTHER RULE/PATTERN-BASED CLASSIFIERS

Decision Trees and CBA: The following two classification algorithms are different from CAEP by effectively using just one rule/pattern to classify an instance. Decision-tree classification algorithms such as C4.5 [168] use a set of rules having disjoint mds learned in a greedy manner. The CBA classification algorithm [122] learns and uses a set of rules where different rules can have non-empty mds intersection; a fixed ordering is imposed on the rules and CBA uses only one rule, namely the first according to the ordering, among the rules whose body matches a given instance to classify the instance. The greedy search used by these algorithms often misses many highly discriminative rules. Using just one rule to classify an instance also has a higher chance of leading to a highly biased decision.

Other Aggregation-based Classification Algorithms: A number of rule-aggregation and pattern-aggregation based classification algorithms have been proposed. They are different from CAEP on how the aggregation is performed, as well as how the rules/patterns are mined and selected. The CMAR algorithm [120] selects, among the set of all matching rules of a given test instance, just one matching rule (based on confidence) to classify the given instance. The CPAR algorithm [226] selects, among the set of all matching rules of a given test instance, some k matching rules (based on their accuracy), and uses the expected accuracy of the k selected rules to classify the given instance. So these methods differ from CAEP with respect to how they aggregate the selected rules. They are also different because they use a very small number of matching rules to classify a test instance. In contrast, CAEP selects lots of EPs where different EPs can have large mds overlap, and it aggregates all matching patterns to classify a given test instance using the CAEPLS score function.

Reference [228] considered using variants of CAEP which differ on how the EPs are mined and selected: the MB-EP approach mines EPs (represented by causal rules) from the feature space of the Markov blanket of the class variable, and the CE-EP approach mines EPs (represented by causal rules) from the feature space involving the direct causes and direct effects of the class variable.

Closely related to CAEP are two classification algorithms, namely iCAEP [231] and JEP-Classifier [113]. Roughly speaking, for a given test instance t and each class C_i, iCAEP first computes a set of EPs $\mathcal{E}_i$, from the EPs of C_i, to cover t, and then it computes a likelihood score for t and C_i using the selected set of EPs using a sum of log of the probabilities associated with those EPs. JEP-Classifier uses only jumping EPs for classification and the likelihood scores it computes are closely related to that of CAEP. Reference [118] combined DeEPs with K-Nearest

Neighbors (KNN) for classification. Reference [68] examined the issue of further improving EP-based classifiers via bagging.

4.8   DISCUSSION

This chapter presented the CAEP approach to classification. The philosophy of the approach is to aggregate the discriminating signals contained in many discriminative patterns that match an instance in order to arrive at an accurate classification. The approach has several notable strengths, including good classification performance in general and on tiny training datasets in particular. CAEP belongs to the families of rule-based and pattern-based classifiers.

CAEP was first proposed in [63]. It was the first pattern-based "aggregation style" classifier. DeEPs was presented in [114].

References [33, 170] provide surveys on pattern/rule-based classification. Reference [83] gives a survey of emerging pattern based approaches for classification. Reference [131] reports a study of the impact of resampling methods for contrast pattern based classifiers in imbalanced data. The following papers studied emerging pattern based classifiers and their properties [70, 71, 82, 84, 134, 145, 152].

A number of papers have studied "associative classification" [1, 23, 201]. These methods make use of association rules by selecting one rule when classifying a test instance.

CHAPTER 5

CAEP for Classification on Tiny Training Datasets, Compound Selection, and Instance Selection

This chapter first shows that CAEP (see Chapter 4) has fairly good empirical performance, and outperforms other classification algorithms, when the training datasets are tiny. It then presents an iterative algorithm for compound selection, which minimizes the amount of human effort in the selection process. The algorithm exploits CAEP's ability to perform well on tiny training data. This chapter also generalizes the compound selection problem to the semi-supervised extreme instance selection problem, and gives an algorithm to solve the problem by adapting the iterative algorithms for compound selection. It also compares the semi-supervised extreme instance selection problem with semi-supervised learning.

Sections 5.1 and 5.2 are based on [18]. The goal of Reference [18] is to develop an effective method for compound selection (for use in drug design) on the basis of only a small number of reference molecules that require "potency" labeling by humans. Such a method is useful for the early stages of a hit-to-lead transition or lead optimization process in medicinal chemistry. Traditional Quantitative Structure-Activity Relationship (QSAR)-based methods use quantitative models based on structure-activity relationships. However, QSAR-based methods require high-quality training datasets containing large numbers of compounds with potency levels, which require a large amount of human effort and are often not feasible when trying to identify novel hits or leads.

5.1 CAEP PERFORMS WELL ON TINY TRAINING DATA

After performing experiments which confirmed that CAEP has high classification accuracy, Reference [18] evaluated CAEP's predictions based on very small training datasets having very few compounds. Specifically, the authors considered several datasets of compounds. For each dataset, they assembled training datasets of 10, 5, or 3 compounds from each of the low-potency and high-potency classes by random selection and predicted the class label of the remaining compounds. Details of the datasets used in [18] are given in Section 5.1.1.

Reference [18] used only minimal jumping emerging patterns in CAEP. Moreover, it used a novel way to normalize the scores for instances: If the numbers of JEPs for the classes are small, the authors normalize the CAEPLSr scores for a class by dividing them by the sum of $sup * GrowthRate$ for all JEPs of the class.

Reference [18] compared CAEP (called ECP in the paper) with binary QSAR [105] and decision trees, predicting test compounds to belong to either the "micromolar" or "nanomolar" potency class (see Section 5.1.1).

Table 5.1 presents the average accuracies (obtained by averaging entries in Table 6 of [18]). Clearly CAEP performed best and its average accuracies are 69.7% or higher. In contrast the other two methods' average accuracies are in the 50–61.3% range. It turns out average accuracies of 69.7% or higher on tiny training data is useful for the compound selection problem discussed next.

Table 5.1: Average accuracies of three classifiers on tiny training data

Training Set Size Per Class	CAEP	BINQSAR	Decision Tree
3	0.697	0.575	0.5
5	0.705	0.568	0.605
10	0.731	0.587	0.613

It is not an accident that CAEP has good performance in these experiments: the good performance happened because (1) CAEP uses aggregation (combination) of all discriminative emerging patterns that exist in the tiny training data, (2) discriminative emerging patterns can be extracted from even a tiny number of compounds, and (3) all possible minimal JEPs are used (instead of subsets of patterns extracted using greedy searches as exemplified by decision-tree algorithms). Importantly, the fact that each emerging pattern is a combination of several simple single-attribute conditions should also have contributed.

5.1.1   DETAILS ON DATA USED FOR COMPOUND SELECTION

Reference [18] considered four publicly available molecular compound datasets. Each dataset was separated into two classes of compounds with potency (measured by IC50) above or below 1 μM, with "nanomolar" denoting "above 1 μM" (the low potency class), and "micromolar" denoting "below 1 μM" (the high potency class). Details of the four publicly available datasets are: benzodiazepines (BZR) [$n = 321, n_h = 283, n_l = 38$], dihydrofolate reductase inhibitors (DHFR) [$n = 586, n_h = 249, n_l = 337$], glycogen synthase kinase-3 inhibitors (GSK3) [$n = 464, n_h = 281, n_l = 183$], and HIV protease inhibitors (HIVPROT) [$n = 967, n_h = 821, n_l = 146$]. Inside the square brackets, n is the total number of compounds, n_h is the number of instances in the high class (micromolar), and n_l is for the low class (nanomolar).

Reference [18] used 61 1D and 2D molecular descriptors (computed using MOE [143]) as features. These descriptors were discretized (each into 10 bins). The entropy-based discretization method eliminated a number of descriptors that were mapped to a single interval and thus were useless for emerging pattern mining. This elimination step substantially reduced the number of descriptors for BZR and DHFR (where 18 and 19 remained, respectively), but the majority of descriptors remained for GSK3 (47) and HIVPROT (45).

5.2   USING CAEP FOR COMPOUND SELECTION

Exploiting CAEP's power on tiny training datasets, Reference [18] presented an iterative algorithm for compound selection, which minimizes the amount of human effort in the selection process.

The setting is as follows: There is a large set of compounds, some of which have potential to become candidates for a drug. The candidate selection process will select high potential candidate compounds, with minimal input and effort by humans. Human effort is concerned with assessing the potency levels of some compounds using experimental means.

Algorithm 5.4 gives the pseudo-code of the iterative algorithm. In essence, the algorithm repeatedly performs the following sequence of operations: draw a small set S of samples from the current set D, ask humans to provide potency levels (e.g., by laboratory experiments) for compounds in S, divide S into two classes using median potency level of S, use S to build a CAEP classifier, use that classifier to classify compounds in D, and remove compounds with low predicted potency from D. The process terminates when only a small number of compounds remain.

Algorithm 5.4 Iterative Compound Selection

Parameters: k (number of compounds per class for building classifiers)
Input: A set D of compounds
    (1) Repeat
        (2) Randomly select a small set S of $2k$ compounds from D
        (3) Get the potency of the compounds in S (labeling by humans)
        (4) Divide S into a high potency and a low potency class,
            using the median potency value as the threshold
        (5) Use S to build a CAEP classifier
        (6) Use the CAEP classifier to classify compounds in $D - S$
        (7) Remove from D all compounds predicted to be "low potency"
    (8) Until $|D|$ is very small

Reference [18] reported the following experiment results: For each activity class and training set size ($k = 5$ or 10), the algorithm consistently reached convergence within the first eight or fewer iterations, producing selected sets of fewer than 10 compounds with average potency in the submicromolar range.

5.3   ITERATIVE ALGORITHM FOR EXTREME INSTANCE SELECTION

The general extreme instance selection problem is the following: given a large set D of data instances, the goal is to identify a small number of instances that are most extreme with respect to some property π. The challenge is that the π values for the instances are not given in D, and human effort is needed to determine π's values. So we want to identify the extreme instances by asking humans to determine the π values for a minimal number of instances.

Algorithm 5.4 can be adapted to solve the general extreme instance selection problem. In fact, we just need to replace the word "compound" in Algorithm 5.4 to "instance," and change "potency" to π. We will call the resulting algorithm "Semi-Supervised Extreme Instance Selection Algorithm." We note that this algorithm can be used to perform semi-supervised ranking for selecting some top-ranked instances; the human experts provide ranking of $2k$ instances in each of the iterations.

5.4   SEMI-SUPERVISED EXTREME INSTANCE SELECTION VS. SEMI-SUPERVISED LEARNING

The extreme instance selection problem is related to, but also somehow different from, semi-supervised learning [235]. This can be explained from two perspectives.

(a) The two problems are similar in that they consider the situations where labeled instances are often difficult, expensive, or time consuming to obtain, as they require the efforts of experienced human workers. They both aim to produce optimal results while minimizing human effort.

(b) The two problems are different due to the following: traditional semi-supervised learning is concerned with learning a model of some sort (namely classification, clustering, regression, ranking). (Reference [235] contains discussions on semi-supervised learning for all those problems listed in the above parenthesis.) In contrast, extreme instance selection is concerned with finding a small number of instances with extreme values for some attribute/property.

Extreme instance selection is perhaps most related to semi-supervised ranking, although the former only wants to rank very few instances whose ranks are the highest. We note that Algorithm 5.5 requires very little human effort, and does not require humans to give ranks to a large number of instances.

Algorithm 5.5 Semi-Supervised Extreme Instance Selection Algorithm

---

Parameters: k (number of instances per class)

Input: A set D of instances

    (1) Repeat

        (2) Randomly select a small set S of $2k$ instances from D

        (3) Get the π values of the instances in S (labeling by humans)

        (4) Divide S into a high-PI and a low-PI class,

            using the median π value as the threshold

        (5) Use S to train a CAEP classifier

        (6) Use the CAEP classifier to classify instances in $D - S$

        (7) Remove all instances from D predicted to be low-PI

    (8) Until $|D|$ is very small

CHAPTER 6

OCLEP: **One-Class Intrusion Detection and Anomaly Detection**

Intrusion detection plays an important role for ensuring the security of various protected systems. Intrusion detection is a special case of anomaly detection, and it is also related to novelty detection and outlier detection. Very often the underlying data for intrusion detection have high dimensions and perhaps also contain many categorical attributes. Many existing intrusion-detection and anomaly-detection approaches directly rely on the use of distance metrics, while some others indirectly rely on distance metrics (e.g., they use models that involve attribute-wise differences between data points). These approaches may perform poorly due to pitfalls of distance metrics in high-dimensional spaces and due to the existence of categorical attributes.

This chapter presents an emerging pattern based method for intrusion detection called OCLEP (One-Class Classification Using Length of Emerging Patterns). As the name suggests, OCLEP is based on the use of the length of jumping emerging patterns. Key properties of OCLEP include the following. (1) It uses only *one-class training data* to build an intrusion-detection system. (2) The constructed intrusion-detection system does not rely on a mathematical model, making it hard for attackers to figure out the details of the detection system. (3) The constructed intrusion-detection system does not use distance metrics, and the method can effectively handle both categorical attributes and numerical attributes. (4) The method has strength in interpretability for analyzing discovered intruders (and anomalies when used for anomaly detection). (5) It is a lazy instance-based approach. In general, OCLEP can be used to perform anomaly detection when the problem is treated as a one-class training problem.

Organizationally, the chapter first gives some general background for intrusion detection and anomaly detection. Then it presents the emerging pattern length-based method OCLEP for intrusion detection. Next it reports experimental evaluation of two variants of OCLEP on two public datasets which have been widely used to evaluate intrusion-detection systems. Finally, some concluding remarks are provided.

6.1   BACKGROUND ON INTRUSION DETECTION, ANOMALY DETECTION, AND OUTLIER DETECTION

Since intrusion detection is a special case of anomaly detection (especially when the training data are all from one class), and since anomaly detection is related to outlier detection, we present some background on intrusion detection, anomaly detection, and outlier detection.

Anomaly detection is concerned with identifying anomalies with respect to a given dataset. Anomaly detection can be treated as a two-class problem and one-class problem. In the two-class case, a dataset containing two classes, representing normal instances and anomaly instances, respectively, is given to train the anomaly-detection system. For the one-class case, a dataset containing just one class (representing the set of normal instances) is given to train the system. One-class anomaly detection is also called *outlier detection*. The one-class case is more challenging. It is more desirable to users as less training data are required and training data for the anomaly class can be hard to obtain, and detection systems based on one-class training data can be deployed earlier in the life-cycle of the application.

For network and system security, anomaly detection includes intrusion detection and masquerader detection. The former is about detecting intruders to a protected system, while the latter is about identifying abusers inside a protected system who use the privileges of others through non-authorized means.

Reference [34] gives a survey of the machine learning and data-mining literature on intrusion detection. It indicates that most methods are classification model-based methods, although clustering-based methods have also been studied. A notable statement from [34] is the following: "Unfortunately, the methods that are the most effective for cyber applications have not been established; and given the richness and complexity of the methods, it is impossible to make one recommendation for each method, based on the type of attack the system is supposed to detect."

Reference [94] provides a survey on outlier detection. From this survey, one can see that most existing outlier-detection approaches rely on the use of distance metrics, while others rely on models that involve the use of data-point differences. Reference [28] points out certain pitfalls of distance metrics in high-dimensional space, asking the question: "When is 'nearest neighbor' meaningful?" Reference [91] contains a discussion of some more recent outlier-detection algorithms. Reference [4] discusses challenges to outlier detection in high-dimensional space.

6.2   OCLEP: EMERGING PATTERN LENGTH-BASED INTRUSION DETECTION

The *length* of a pattern is defined as the number of items (or single-attribute conditions) contained in the pattern. For example, the length of $\{a, c, e\}$ is 3.

Below we use N to denote a training dataset of normal instances.

6.2.1   AN OBSERVATION ON EMERGING PATTERN'S LENGTH

Normally one would use emerging patterns to differentiate instances in different classes. It turns out that one can also use emerging patterns to differentiate instances that are from a common class. Analyzing what we observe from these uses, one would likely see the following difference:

> The emerging patterns differentiating instances of different classes are often short, whereas the emerging patterns differentiating instances of a common class are often long. The former are typically much shorter than the latter.

Experiments [41] on the Mushroom data from the UC Irvine Repository confirmed this observation: the average length of minimal jumping emerging patterns differentiating instances of a common class is 7.78, and the average length of minimal jumping emerging patterns differentiating instances of different classes is 3.03. The experiments involved repeated mining of minimal jumping emerging patterns that occur in an instance but not occurring in a small random set of several hundreds of instances. For differentiating instances of a common class, both the instance and the random set are from the common class. For differentiating instances of different classes, the instance is from one class and the random set is from a different class.

To get a rough idea of what the above observation means, consider this example. To differentiate residential buildings and agricultural crops, one would use emerging patterns of length 1, for example, crops grow but buildings do not. To differentiate one residential building from a collection of other residential buildings, one may need to use emerging patterns of length 3 or larger. This is due to the fact that each single-attribute condition satisfied by one residential building is also satisfied by many other residential buildings; only combinations of several single-attribute conditions can indicate the uniqueness of a particular residential building.

Our intrusion-detection method below makes use of this length difference.

6.2.2   WHAT EMERGING PATTERNS TO USE AND THEIR MINING

To use the observation above in an emerging pattern length-based intrusion-detection method, three issues need to be resolved:

(a) What kinds of emerging patterns should be used?

(b) Should we use the eager approach or lazy approach to mine emerging patterns?

(c) How can we mine emerging patterns in order to increase robustness and efficiency of the intrusion-detection system?

Recall that we consider one-class training-based intrusion detection—we only have one training dataset N representing the set of normal instances. Moreover, the number of training instances can be very large (e.g., > 10,000).

The discussion in Section 6.2.1 already hinted that we need to differentiate instances of a common class as well as differentiating instances of different classes. In particular, when predicting if a new instance is an intruder, we need to mine emerging patterns that differentiate this instance from the normal instances. In the training process, we must utilize the training data to gather needed information for use by the system to predict on new instances. Combining these, we see that we need to mine emerging patterns that differentiate an instance from a set of instances. So one possible (and our chosen) answer to Question (a) is: use jumping emerging patterns that occur in one particular instance but not in a set of normal instances.

Because of the above choice and also because of our decision for Question (c), we need to use the lazy learning approach, where we mine patterns after we see the test instance. Moreover, in this lazy approach we only want to use statistics obtained in the training process, and we do not want to store/use all the emerging patterns mined in that process. (There are too many such patterns.)

For efficiency reasons, the emerging patterns to be used should be computed quickly to make the intrusion-detection system efficient to use. It is also desirable to find ways to increase the robustness of the system. Our answer to Question (c) is to use the BorderDiff algorithm [56] repeatedly in a 1-vs-m fashion: one instance t vs. a random sample M from N of some suitable size m. This helps improve on both robustness and efficiency. By having m be a reasonably small number, we get efficiency. By finding jumping emerging patterns that occur in t but not in M for multiple M, the corresponding derived statistics on the patterns becomes more stable.

We note that jumping emerging patterns mined from one particular "t vs. M" may not be jumping emerging patterns for "t vs. N."

6.2.3   OCLEP'S TRAINING AND TESTING ALGORITHMS

Two versions of OCLEP have been introduced, which mainly differ on which length statistics is used: the original OCLEP method uses average length, whereas OCLEP+ uses minimal length. We present OCLEP+ below.

Given a set $\mathcal{P}$ of mined emerging patterns, let $avgLength(\mathcal{P})$ denote the average length, and $minLength(\mathcal{P})$ denote the minimal length, of the patterns in $\mathcal{P}$.

The training algorithm for OCLEP+ will derive just one value: a cutoff value for the minimal lengths. For this, k instances $t1, \ldots, tk$ are randomly drawn from N; each will contribute a minimal length value. For each tj, r random samples M_i, each containing m instances, are drawn randomly from $N - \{tj\}$. The minimal jumping emerging patterns from $\mathsf{BorderDiff}(tj, M_1), \ldots, \mathsf{BorderDiff}(tj, M_r)$ are collected and used to compute a minimal length ml_{tj}. Having r invocations of BorderDiff on tj vs. r different M_i's helps improve the performance and robustness of the detection system. The list of the k minimal lengths from the k tj's are sorted to yield the cutoff value at the pth percentile.

The suitable size m is determined as follows: if N is large, we choose m in the range of $[200, 800]$. If N is not very large, we choose m to be $|N| - 1$. Choosing a larger m allows us to use more instances as background data to be compared against t, but more computational time is needed. Using m larger than 800 typically does not lead to additional discriminative information but leads to longer computation time. If m is roughly N then the detection system may not

---

Algorithm 6.6 OCLEP+ Training

Parameters: k (size of T), m (size of M), r (number of M per t),
    p (percentile for cutoff value)
Input: N (dataset of normal instances)
Output: A cutoff value κ
    (1) Pick a random subset $T = \{t1, \ldots, tk\}$ of size k from N
    (2) For each $t \in T$ do
        (3) For each $i \in \{1, \ldots, r\}$ do
            (4) Pick a random subset M_i of size m from $N - \{t\}$
            (5) Let PS_i be the set of minimal jumping emerging patterns
                computed by BorderDiff(t, M_i)
        (6) End
        (7) Let $ml_t = \min(\{|P| \mid P \in \cup_{i=1}^{r} PS_i)$
    (8) End
    (9) Sort the list $ml_{t1}, \ldots, ml_{tk}$ in decreasing order
    (10) Return the cutoff value κ at p-percentile of the sorted list

---

Algorithm 6.7 OCLEP+ Testing

Parameters: m (size of M), r (number of M per t), κ (the cutoff value)
Input: N (dataset of normal instances), x (test instance)
Output: Decision whether x normal or intruder.
    (1) For each $i \in \{1, \ldots, r\}$ do
        (2) Pick a random subset M_i of size m from N
        (3) Let PS_i be the minimal jumping emerging patterns
            computed by BorderDiff(x, M_i)
    (4) End
    (5) Let $ml_x = \min(\{|P| \mid P \in \cup_{i=1}^{r} PS_i)$
    (6) If $ml_x > \kappa$ then classify x as normal
    (7) Else classify x as intruder

---

be as robust (compared against a medium m), as only one set of jumping emerging patterns is available. Having m in the hundreds range usually offers good speed-information tradeoff and better robustness. In experiments on the NSL-KDD data from the Canadian Institute for Cybersecurity, parameters for OCLEP+ were set as follows: $k = 800$, $m = 400$, $r = 7$, and $p = 95\%$.

6.3   EXPERIMENTAL EVALUATION OF OCLEP

Experiments to evaluate OCLEP were conducted on two public datasets related to security, one about intrusion detection and the other about masquerader detection. The discussion below focuses on the former, with a brief description of the latter.

6.3.1   DETAILS OF THE NSL-KDD DATASET

The NSL-KDD dataset[1] was used to evaluate OCLEP+'s performance. It is an improved version of the KDDCUP dataset;[2] the modifications were performed to remove certain design deficiencies of the KDDCUP data [195]. KDDCUP was used for the Third International Knowledge Discovery and Data Mining Tools Competition, which was held in conjunction with KDD-99; the dataset includes a wide variety of intrusions that are simulated in a military network environment. Both KDDCUP and NSL-KDD have been widely used in evaluating intrusion-detection algorithms.

The instances of NSL-KDD have 41 features. The NSL-KDD dataset includes a file (KDDTrain+_20Percent) that contains both anomaly as well as normal instances. For one-class training for OCLEP+, the anomaly instances were removed from this file, yielding a one-class training file with 13,499 normal instances. The KDDTest+ file of NSL-KDD containing 22,544 instances (both normal and abnormal) was used to evaluate the classifiers. The equi-width discretization method was used on the numerical attributes. Observe that in one-class training we do not have access to the classes. Parameters for OCLEP+ were set as follows: $k = 800$, $m = 400$, $r = 7$, and $p = 95\%$.

6.3.2   INTRUSION DETECTION ON THE NSL-KDD DATASET

In general, experiments found that OCLEP+ achieved better results than competing methods on the NSL-KDD dataset. The results are summarized in Tables 6.1 and 6.2. The tables compare performance of OCLEP+, OCLEP, and one-class SVM (from [39]) with three types of kernels. Table 6.1 gives raw counts for the confusion table, whereas Table 6.2 gives results on several accuracy measures.

6.3.3   MASQUERADER DETECTION ON COMMAND SEQUENCES

In masquerader attacks, an intruder uses another person's identity to do things inside a protected system. A common approach to detect masqueraders is to compare a user's recent behavior against his/her profile of typical behavior and to use deviation as indication of masquerading.

The dataset[3] provided by [67] was used. It consists of sequences of "truncated" commands for 50 users; each user is represented by a sequence of 15,000 commands. The first 5,000 com-

---

[1]NSL-KDD: http://www.unb.ca/cic/research/datasets/nsl.html.
[2]KDDCUP: https://archive.ics.uci.edu/ml/machine-learning-databases/kddcup99-mld/kddcup99.html.
[3]http://www.schonlau.net

Table 6.1: Performance comparison: confusion table

Method	True Positive	False Positive	True Negative	False Negative
OCLEP+	9810	1358	8353	3023
OCLEP	9762	1724	7987	3071
OneClass SVM: Linear	10615	4593	5118	2218
OneClass SVM: Poly	10859	4661	5050	1974
OneClass SVM: RBF	12825	9706	5	8

Table 6.2: Performance comparison on accuracy measures

Method	Precision	Recall	F-score	Accuracy
OCLEP+	87.84	76.44	81.75	80.57
OCLEP	84.99	76.07	80.28	78.73
OneClass SVM: Linear	69.80	82.72	75.71	69.79
OneClass SVM: Poly	69.97	84.62	76.60	70.57
OneClass SVM: RBF	56.92	99.94	72.53	56.91

mands of each user are "clean data" (i.e., legitimately issued by the user), and the last 10,000 commands were probabilistically injected with commands issued by 20 users outside the community of the 50. The commands are grouped into blocks of 100 commands. The commands in one block are either all clean, or all masquerade attacks (a "dirty block"). The task is to accurately classify the user-command blocks into two categories: self (i.e., the clean blocks) and masqueraders (i.e., the dirty blocks).

In several one-class training settings, the original OCLEP algorithm (using average length) achieved results (on area under the curve of receiver operating characteristic curves) comparable to that for one-class support vector machine (SVM). Each command block (a sequence) was transformed into a feature vector of some type in the experiments. For details see [41, 42].

6.4 DISCUSSION

This chapter presented a one-class intrusion-detection method called OCLEP, with a focus on OCLEP+, which is based on the use of minimal lengths of jumping emerging patterns. Experiments show that OCLEP+ performs better than competing methods such as one-class SVM. More details on the experiments can be found in [41, 42, 60]. It should be noted that the method can be used for anomaly detection, outlier detection, and novelty detection, especially when the problems are treated as one-class problems.

The good performance of the OCLEP methods indicate that emerging patterns can be used as substitute for distance metrics for anomaly/outlier detection. The OCLEP algorithms also have strength in interpretability of anomaies/outliers, which is an important issue for anomalies/outlier detection [192]. Indeed, the minimal-length jumping emerging patterns found for a given instance predicted to be an anomaly/outlier can help explain in what ways the instance is an anomaly/outlier.

Some other researchers also used emerging patterns to perform anomaly detection. Reference [37] used emerging patterns in anomaly detection. Reference [30] used emerging patterns in anomaly detection for an application in the setting of service-oriented architecture; it found that combining emerging patterns with the likelihood score of CAEP (plus two filters, namely "patterns whose likelihood score is lower than a threshold are not used" and "instances without a predicted class is assume to be normal") produced the best performance under a number of settings.

CHAPTER 7

CPCQ: **Contrast Pattern Based Clustering-Quality Evaluation**

This chapter presents a contrast pattern based method called CPCQ for clustering-quality evaluation. Special strengths of the method include (1) it does not use distance metrics, and (2) the underlying algorithm produces, for each cluster, a number of contrast patterns to represent the uniqueness of the cluster.

Organizationally, this chapter starts with some background material on clustering-quality evaluation. Then it gives the rationale for CPCQ, followed by techniques to measure the quality of individual CPs (contrast patterns) and to measure the diversity of (sets of) high-quality CPs. Then it defines the CPCQ measure and presents a method to compute some best groups of CPs to maximize CPCQ values. This is followed by discussion on experimental evaluation of CPCQ. Finally, some concluding remarks are given.

This chapter uses the term CPs instead of "emerging patterns" since the papers [125, 126] presenting CPCQ used the former. The two terms have been used as synonyms.

This chapter will use the following notations and assumptions. Let D denote a given dataset having only categorical attributes,[1] and let $\mathcal{C} = \{C_1, \ldots, C_r\}$ denote a clustering of D (each C_i denoting a cluster). We will use one shortest pattern P in MGS to represent an equivalence class $< MGS, MP >$ of CPs, where MGS denotes the set of minimal generators and MP denotes the closed pattern for the equivalence class. Below when we discuss CPs, we refer to the CPs that represent the equivalence classes containing these CPs.

7.1 BACKGROUND ON CLUSTERING-QUALITY EVALUATION

As clustering[2] is a subjective process in nature [221], there is no clustering-quality index that is universally accepted as the best quality index. Due to the popularity and usefulness of clustering analysis, many clustering-quality indices have been proposed.

Existing clustering-quality indices can be divided into three categories: pairwise-distance based, entropy-based, and frequent-item based.[3]

---

[1]This chapter's methods can be applied to numerical datasets after they are discretized.
[2]Background material on clustering is given in Chapter 8.
[3]Formal definitions of these indices are omitted here; they can be found in [125, 126].

- In the pairwise-distance-based category, Dunn's index is defined as the ratio of the minimal inter-cluster distance to the maximal intra-cluster distance; other distance-based indices include the Davies–Bouldin index, the Silhouette index, the Hamming diameter index, and the Hamming radius index.

- The entropy-based clustering-quality index employs the well-known concept of entropy. It emphasizes intra-cluster purity, and it prefers clusterings in which many of the items which occur in any given cluster are frequent items in the cluster. An item is said to be frequent in a cluster if its support in the cluster is higher than a given support threshold. (It should be noted that this entropy measure does not have anything to do with classes.)

- The frequent-item-based clustering-quality index is based on the notion of frequent items. A clustering is considered good if many items of any given cluster are frequent within that cluster and there is little overlap of frequent items across clusters.

Like the contrast-pattern-based index CPCQ, the above are all internal validation measures in the sense that they only rely on the data and do not rely on information outside of the data (such as known classes).

7.2   CPCQ'S RATIONALE

For clustering-quality evaluation, one can view the clusters as classes. In particular, one can mine contrast patterns (CPs) distinguishing the clusters, and use the quality of the CPs to evaluate the quality of the clustering.

The design of the CPCQ measure followed this rationale:

A high-quality clustering, capturing natural concepts in the corresponding data, should have many *diversified, high-quality* CPs contrasting the clusters.

Following this rationale, CPCQ combines two factors into one numerical value: the quality of the CPs and the diversity (difference) among the CPs. The measure's value will be high if there are many high-quality CPs and there are many highly different CPs among the high-quality CPs.

7.3   MEASURING QUALITY OF CPS

References [125, 126] used experimental results on well-studied datasets with classes to identify and validate what should be used to measure quality of CPs. These experiment results include those reported in Table 7.1 on the Mushroom[4] dataset.

In Table 7.1, Column 2 is the average length of the closed patterns, Column 3 is the average length of the MGs, Column 4 is the length ratio (see below), and Column 5 is the

---

[4]From the UCI Repository. The dataset has 8,124 tuples, 22 attributes, and two classes (edible, poisonous).

Table 7.1: Values of measures on permuted mushroom. x: percent of data exchanged between the two classes

x	Average Length Closed CP	Average Length MG CP	Average lenRatio	Average $supp$
0	12.70	3.21	3.96	294
0.1	12.90	3.52	3.66	226
1	14.42	4.24	3.40	95
10	15.10	4.67	3.23	49

average support of all frequent equivalence classes of CPs.[5] Column 1 gives x values, which indicate what percent of data exchanges between the classes happened. For each x, $x\%$ of the instances of the edible class are re-labeled as poisonous and $x\%$ of the instances of the poisonous class are re-labeled as edible. These exchanges are used to create different versions of Mushroom with varying quality. Clearly larger x leads to a version where the quality of the classes is less desirable and hence the version is of lower quality.

For each equivalence class, lenRatio is defined to be the length of the closed CP divided by the minimal length of the MGs.

Table 7.1 indicates that lenRatio and $supp$ both decrease, while the other two length-related columns both increase, as the version becomes worse. (Note that lenRatio combines the other two length-related columns.) Moreover, lenRatio and $supp$ are not correlated. References [125, 126] selected lenRatio and $supp$ as indicators of high-quality CPs; larger is better for both.

Suppose $< MGS, MP >$ is an equivalence class and P is a shortest pattern in MGS representing the equivalence class. The *quality* of pattern P is measured by $QC(P) = supp(P) \times |MP|/|P|$.

From a pure clustering perspective, one can easily see that we should prefer long closed CPs P as all instances in $mds(P)$ share the values of P, increasing the coherence of these instances because they are in a common cluster. Moreover, CPs having larger support also help increase coherence of their home clusters. On the other hand, in Chapter 6 it was argued that shorter MGs indicates it is easier to distinguish the clusters. So we also should desire shorter MGs.

---

[5] Recall that all CPs (which are patterns) can be divided into equivalence classes, and each equivalence class is described by a pair $< L, R >$, where L is a set containing all minimal generator patterns and R is a closed pattern, for the equivalence class (see Chapter 2).

7.4   MEASURING DIVERSITY OF HIGH-QUALITY CPS

Different from the QC measure of patterns, the diversity of high-quality CPs for a clustering C should reflect the amount of difference in the high-quality CPs that exist for C. The diversity measure should be concerned with two kinds of differences, one on matching data instances of the CPs and the other on items (namely attribute-value and attribute-interval pairs) used in the CPs. For matching data, higher diversity is indicated by smaller mds overlap among CPs. For items, higher diversity is indicated by smaller overlap in the items used among CPs.

There is another issue: Due to the large number of CPs that typically exist, it is hard to compute mds-based difference and item-based difference on all patterns. It turns out that this issue has impact on the effectiveness of the CPCQ measure, in addition to computational efficiency.

The solution taken in [125, 126] is to compute the diversity of some optimal set of CPs, which consists of a small number of groups.

Intuitively, each group represents one way to describe the differences among the clusters. Hence, with each group, the difference in the CPs is evaluated on both mds and items. Between groups, the difference in the CPs is evaluated only on items, as the patterns in each group are expected to match (nearly) all data instances.

Technically, to formalize the "many diversified" aspect of the CPCQ measure, we need to define overlap between patterns, overlap among patterns in a group, and overlap among patterns between groups. Overlap can be in terms of items, or in terms of mds. A *group* of CPs is a finite set of CPs; each in the set represents its respective equivalence class.

There are three variants for item overlap. The *item overlap* between two CPs P_1 and P_2 is defined as $ovi(P_1, P_2) = |P_1 \cap P2|$ (the number of shared items). The item overlap of a group G of CPs is defined as $ovi(G) = avg\{ovi(P_1, P_2) \mid P_1 \in G, P_2 \in G, P1 \neq P2\}$ (the average item overlap between CPs of G). The item overlap between two groups G_1 and G_2 of CPs is defined as $ovi(G_1, G_2) = avg\{ovi(P_1, P_2) \mid P_1 \in G_1, P_2 \in G_2\}$ (the average item overlap between two groups G_1 and G_2 of CPs).

Similarly, two variants of mds overlap can be defined. The mds overlap of two CPs P_1 and P_2 of a common home[6] cluster C is defined as $ovt(P_1, P_2, C) = |\text{mds}(P_1, C) \cap \text{mds}(P_2, C)|$ (the number of tuples in C that match both P_1 and P_2). The mds overlap of a group G of CPs is defined as $ovt(G) = avg\{ovt(P_1, P_2) \mid P_1 \in G, P_2 \in G, P_1 \neq P_2\}$. For a given cluster, since each group of CPs is expected to match the entire cluster, we do not consider the mds overlap between groups.

7.5   DEFINING CPCQ

We can now define the quality of a cluster C in terms of groups of CPs. Two scenarios exist: (a) In the single-group situation, given a group G of CPs for a cluster C, the CP-based quality

---

[6]The home class (data group) of an EP/CP is the class where the pattern has larger support.

of cluster C with respect to G, denoted by $QC_G(C)$, is defined as

$$QC_G(C) = \frac{\sum_{P \in G} QC(P)}{(1 + ovt(G)) \times (1 + ovi(G))}.$$

(b) In the situation of multiple groups, given N groups $G_1, \ldots, G_N$ of CPs for a cluster C, the CP-based quality of cluster C with respect to the N groups, denoted by $QC_{G_1,\ldots,G_N}(C)$, is defined as

$$QC_{G_1,\ldots,G_N}(C) = \frac{\sum_{i=1}^{N} QC_{G_i}(C)}{1 + avg_{1 \leq i < j \leq N} ovi(G_i, G_j)}.$$

We are now ready to define the CPCQ quality measure of categorical data clusterings. Suppose $\mathcal{C} = \{C_1, C_2, \ldots, C_r\}$ is a clustering of a categorical dataset D and suppose that we have selected (through some practical and efficient method) N groups of CPs for each of the r clusters. The CPCQ index, denoted by $QC(\mathcal{C})$, is defined as the size-weighted average of the quality of the clusters, that is,

$$\text{CPCQ}(\mathcal{C}) = \sum_{i=1}^{r} \frac{|C_i|}{|D|} QC_{G_1,\ldots,G_N}(C_i).$$

7.6 MINING CPS AND COMPUTING THE BEST N GROUPS OF CPS TO MAXIMIZE CPCQ VALUES

Deriving an accurate CPCQ index value depends on identification of the most representative, high-quality groups of CPs. However, there are often a large number of CPs, and hence a large number of possible groups. It is thus a challenge to efficiently select the top N high-quality groups of CPs. The difficulty is compounded by the requirement that each group of CPs should have little within-group item and mds overlap and should have little between-group item overlap with other groups.

To mine the CPs and compute the CPCQ value of a clustering, References [125, 126] use a heuristic search algorithm which proceeds in two main steps.

(a) The first step mines all the CPs and their equivalence classes from the clusters of the given clustering. For example, the DPMiner algorithm [116] can be used. For this, there is a need to set the minimum support thresholds for the clusters.

(b) The second main step constructs the top N highest quality groups of CPs and then computes the CPCQ value based on those groups.

The algorithm constructs the top N groups incrementally for each cluster of the given clustering. After computing $i - 1$ groups, it builds the ith group by adding one CP at a time, iteratively selecting a CP which, when added to the ith group, results in the greatest possible quality improvement for the groups, then adding this CP to the ith group. To give the maximum improvement, the selected CP should have large support and large lenRatio, low mds overlap, and low item overlap with all the other selected CPs, both in its own group and in the other existing groups. When the addition of new CPs results in no further improvement, the expansion of the ith group is stopped and the construction of the next group is started.

7.7 EXPERIMENTAL EVALUATION OF CPCQ

References [125, 126] reported various experiment results on several datasets (including Mushroom and Splice-junction) from the UCI Repository evaluating the usefulness of the CPCQ index. More specifically, it used several clustering algorithms, as well as different parameter settings, to produce different clusterings, and compared the performance of various clustering-quality indices on those clusterings as well as the expert-given classes (the original classes of the datasets). A quality index is considered strong if it can identify the clustering corresponding to the expert-given classes as the best clustering. It found that CPCQ often identifies the expert-given classes as the best clustering. Moreover, it found that many other clustering-quality indices often give higher quality values to clusterings with larger numbers of clusters, but this is not the case for CPCQ.

7.8 DISCUSSION

This chapter presented a contrast pattern based, internal, clustering quality measure called CPCQ. The CPCQ index was designed based on the rationale that a high-quality clustering should have a diversified set of many high-quality contrast patterns distinguishing its clusters. It presented mathematical formulae to formalize the intuitive ideas behind this rationale. It also summarized experimental findings for evaluating the performance of CPCQ.

This chapter is related to Chapter 8, which presents a clustering algorithm aimed to produce clusterings that maximize the CPCQ values. The study leading to CPCQ was first motivated by the success of OCLEP (see Chapter 6) in general, and by the usefulness of shorter emerging patterns for anomaly/outlier detection. Together the three chapters (6, 7, 8) demonstrate that characteristics of emerging/contrast patterns can be used to indicate result quality of data-mining and machine learning tasks.

CPC: **Pattern–Based Clustering Maximizing** CPCQ

This chapter presents a pattern-based clustering algorithm called CPC, together with some experiments on textual blog data, to illustrate CPC and its strength. The CPC algorithm works with frequent patterns mined from given data to produce clusterings aimed at maximizing CPCQ. So CPC is aimed at maximizing the quality and diversities of contrast patterns that distinguish the produced clusters.

Special strengths of CPC include: (1) CPC does not use distance metrics, and (2) the method produces interpretable results by associating each cluster with a set of contrast patterns characterizing it. Moreover, the method focuses its attention on the quality of the contrast patterns associated with the clustering results, instead of the usual quality metrics that are often used by most traditional algorithms.

Organizationally, the chapter starts with some background material on clustering analysis. Then it presents the pattern-based clustering algorithm CPC. Next it describes results of experiments on textual blog data which also illustrate what CPC produces. Finally, some concluding remarks are given.

8.1 NOTATIONS

Below we use the term "contrast patterns" (CPs) instead of "emerging patterns" since the papers presenting these algorithms used the former. The two terms have been used as synonyms. Importantly, CPC views equivalence classes, rather than frequent patterns, as basic units of patterns. That is, a (contrast) pattern below refers to the equivalence class of patterns that contains that pattern. This is done both for computational efficiency and for the reason that patterns that have the same matching datasets should be considered as having the same "behavior/meaning."

Given a pattern P, let $\mathsf{mgLen}(P)$ denote the average length of the minimal-generator patterns in the equivalence class of P, and let $P_{\max}$ denote the closed pattern in the equivalence class of P.

8.2   BACKGROUND ON CLUSTERING AND CLUSTERING EVALUATION

Clustering is concerned with partitioning a given dataset into a certain number of clusters (groups, subsets, or categories), such that instances in the same cluster are very similar to each other and instances in different clusters are very different from each other. Clustering can be applied to relational data and non-relational data. The data can have high dimensionality as well as large numbers of instances. Moreover, the meaning of "similarity between data instances" is not fixed; different algorithms often work with different measures of similarity and use their own measures to produce their ideal clustering results.

Cluster analysis is an important tool for explorative analysis on unlabeled data, which is often used to gain some understanding of the intrinsic structures of the data. As a result, a huge number of algorithms have been introduced. References [220, 221] are two fairly recent surveys on clustering.

Clustering is a subjective process in nature [221]. In particular, there is no universally accepted definition of similarity. Similarity is often measured by a distance metric, selected out of a large number of possible distance metrics. For high-dimensional data, there exist serious pitfalls of distance metrics; Reference [28] asked the question: "When is 'nearest neighbor' meaningful?" The fact that the understanding on the given data for clustering is not deep (since the goal of performing clustering is often exploratory analysis) also implies that it is hard to select and define the optimal distance metric. We note that similarity does not need to be defined by a distance metric.

8.3   PROBLEM SETTING AND GUIDING IDEAS FOR CPC

The CPC algorithm can perform clustering on any dataset; if the dataset contains numerical attributes, then some appropriate preprocessing is needed so that patterns can be mined. It relies only on frequent patterns in the clustering process. It does not make use of distance metrics. The CPC algorithm also produces characterizing summaries of the clusters using small sets of contrast patterns (CPs).

The CPC algorithm was designed to form clusterings with the aim of maximizing the CPCQ value (see Chapter 7). As a CP of a clustering is a pattern appearing with significantly larger frequency in its home cluster than in any other clusters, a CP is a highly discriminative pattern describing its home cluster and distinguishing that cluster from other clusters. As CPCQ measures the quality, abundance, and diversity of the CPs of the clustering, the CPC algorithm will produce a clustering so as to maximize the number and diversity of high-quality CPs associated with the clustering.

The guiding idea of the CPC algorithm is: when identifying frequent patterns to make them high-quality contrast patterns of some clusters, the decision should be based on how many other frequent patterns will also become high-quality contrast patterns of those clusters. It should be emphasized that all such evaluations and decisions will be only based on characteristics of frequent patterns.

8.4   MAIN TECHNICAL MEASURES

CPC does not directly use CPCQ. Instead, CPC uses a coherence measure that is mostly based on the matching datasets of patterns to guide the clustering process. This coherence measure, termed *Mutual Pattern Quality* (MPQ), can be viewed as a distance function on a pair of patterns. MPQ essentially measures the number (richness) and quality of *other* patterns that may become high-quality CPs when the two patterns (of the given pair) become CPs of the same cluster. A high MPQ value indicates that the two patterns should belong to the same cluster, while a low value indicates they should belong to different clusters.

There are two variants of the MPQ measure, one between two patterns, and the other between a pattern and a set of patterns.

8.4.1   MPQ BETWEEN TWO PATTERNS

The MPQ formula makes use of two other formulae, which are respectively called PQ2 and PQ1. PQ2 is basically the raw MPQ and PQ1 is used to normalize PQ2.

(a) Given patterns P_1 and P_2, we use $PQ2(P_1, P_2)$ to reflect the desirability of making P_1 and P_2 CPs of a common cluster. The desirability is proportional to the richness and quality of other new CPs X of that common cluster that will result if we merge P_1 and P_2 into the same cluster. Based on how CPCQ was defined in Chapter 7, we see that the quality is determined by the length ratio, the support, and the growthRate of all possible patterns X.

PQ2 is defined as follows:

$$PQ2(P_1, P_2) = \sum_X \left(\frac{|\mathsf{mds}(P_1) \cap \mathsf{mds}(X)| * |\mathsf{mds}(P_2) \cap \mathsf{mds}(X)|}{|\mathsf{mds}(X)|} * \left(\frac{|X_{\max}|}{\mathsf{mgLen}(X)|} \right)^2 \right).$$

In the summation, X ranges over all patterns. We explain the rationale below.

First, the PQ2 formula includes a part to evaluate the length ratio of X. Based on experimental findings, the formula uses the square instead of the linear form of the length ratio. Second, the part

$$\frac{|\mathsf{mds}(P_1) \cap \mathsf{mds}(X)| * |\mathsf{mds}(P_2) \cap \mathsf{mds}(X)|}{|\mathsf{mds}(X)|}$$

mostly reflects how much of $\mathsf{mds}(X)$ is contained in $\mathsf{mds}(P_1) \cup \mathsf{mds}(P_2)$, which indirectly measures the growthRate of X as a CP of the cluster of P_1 and P_2. More specifically, (a) if $\mathsf{mds}(X)$ is fully contained in $\mathsf{mds}(P_1) \cup \mathsf{mds}(P_2)$, then X becomes a jumping emerging pattern for the cluster of P_1 and P_2; (b) if $\mathsf{mds}(X)$ has no overlap with $\mathsf{mds}(P_1) \cup \mathsf{mds}(P_2)$, then X is not a CP for the cluster of P_1 and P_2. (c) Suppose now that $\mathsf{mds}(X)$ is partly but not totally contained in $\mathsf{mds}(P_1) \cup \mathsf{mds}(P_2)$. Let IN denote $\mathsf{mds}(X) \cap (\mathsf{mds}(P_1) \cup \mathsf{mds}(P_2))$ and OUT denote $\mathsf{mds}(X) - (\mathsf{mds}(P_1) \cup \mathsf{mds}(P_2))$. If $|IN| > |OUT|$, then X will be a CP with $\mathsf{growthRate} > 1$, and the quality of X will depend on $|IN| - |OUT|$. On the other hand, if $|IN| < |OUT|$, then X will not be a CP of the cluster of P_1 and P_2.

(b) We now turn to PQ1, which is used to normalize $PQ2$ to define MPQ. PQ1 is aimed at reflecting the richness of patterns X that are already high-quality CPs of the cluster of P_1 or the cluster of P_2 (so they are high-quality CPs associated with P_1 or P_2, before merging P_1 and P_2). This is motivated by the fact that sometimes the reason a pattern X becomes a high-quality CP of the cluster having P_1 and P_2 as CPs is because of the relationship between X and P_1 or between X and P_2. We want to adjust for the influence by that kind of X in the final MPQ formula. PQ1 is defined as:

$$PQ1(Q) = \sum_P |\mathsf{mds}(P) \cap \mathsf{mds}(Q)| \left(\frac{|P_{\max}|}{\mathsf{mgLen}(P)} \right)^2.$$

In this formula, P ranges over all possible patterns; the weight given to P increases with its length ratio and its overlap with $\mathsf{mds}(Q)$, reflecting its potential to contribute to a PQ2 value for Q and any other pattern.

We now turn to the MPQ formula itself. Given a pair of patterns P_1 and P_2, their MPQ value is defined as

$$MPQ(P_1, P_2) = \frac{PQ2(P_1, P_2)}{PQ1(P_1) * PQ1(P_2)}.$$

8.4.2   MPQ BETWEEN A PATTERN AND A PATTERN SET

When adding a pattern P to a (partial) cluster, we need to consider the relationship between P and the cluster. At this time we have selected some set PS of patterns for inclusion as CPs of the cluster; we may still want to add more CPs to the cluster. We represent the cluster by the union of the mds of the patterns assigned to the cluster, that is, $\mathsf{mds}(PS) = \cup_{P \in PS}\mathsf{mds}(P)$. We define three counter parts of the three formulae in Section 8.4.1:

$$PQ2(P, PS) = \sum_X \left(\frac{|\mathsf{mds}(P) \cap \mathsf{mds}(X)| * |\mathsf{mds}(PS) \cap \mathsf{mds}(X)|}{|\mathsf{mds}(X)|} * \left(\frac{|X_{\max}|}{\mathsf{mgLen}(X)} \right)^2 \right),$$

where X ranges over all patterns not in $PS \cup \{P\}$;

$$PQ1(PS) = \sum_P |\mathsf{mds}(P) \cap \mathsf{mds}(PS)| \left(\frac{|P_{\max}|}{\mathsf{mgLen}(P)} \right)^2;$$

$$MPQ(P, PS) = \frac{PQ2(P, PS)}{PQ1(P) * PQ1(PS)}.$$

8.5   THE CPC ALGORITHM

Algorithm 8.8 The CPC Algorithm

Parameters: k (number of clusters), *minSup* (for frequent patterns)

Input: D (dataset) and $\mathcal{P}$ (frequent patterns)

(1) Find weakly related seed patterns to initially define the clusters;

(2) Iteratively add patterns to clusters, selecting patterns with high MPQ values;

(3) Assign remaining patterns as CPs to clusters based on mds overlap;

(4) Assign tuples to clusters based on the CPs they match.

CPC's pseudo-code is given in Algorithm 8.8. We now give more details.

Step 1. Find Seeds by MPQ Minimization: to initialize the k clusters, we define a set of seed patterns as a set of k patterns where the maximum MPQ value between pairs of patterns in the set is very low. Exhaustively searching each possible set is too expensive, so a heuristic is used. Roughly speaking, some M seed sets are generated at random, the best N (for some fixed N) are selected for refinement, and the best refined set is returned. (Tie breaking selects the set having the highest total support.) To refine a set, the seed pair responsible for the maximum MPQ value is targeted; a best replacement P is found for each of these two seeds in the pair; and the seed having the better improvement is replaced by its respective P. This refinement repeats until no improvement is found. Reference [80] include more details on how to select random seeds, M, N, and so on, to shorten the computation time.

Step 2. Assign Patterns by MPQ Maximization: in Step 1, the CP group G of each cluster C_i, denoted by $G(C_i)$, was initialized to contain one of the seed patterns. Step 2 adds strongly related patterns P to CP groups $G(C_1), \ldots, G(C_k)$ by repeatedly searching for the $(P, G(C_i))$ pair that maximizes $MPQ(P, G(C_i))$ and adding P to $G(C_i)$. For efficiency, only patterns P for which $|\text{mds}(P) \cap \text{mds}(G(C_i))|$ is small are candidates in this step. Patterns are added until no such candidate pattern exists.

Step 3. Assign Patterns by PCM Maximization: after Step 2, each cluster only has been assigned a very small number of CPs. A vast number of potential CPs have not been assigned. Moreover, the CP groups created in Step 2 are unlikely to cover the entire dataset under consideration. Step 3 therefore assigns patterns based on their tuple overlaps with each CP group. Together with the CPs already assigned, these CPs typically cover the entire dataset and allow each tuple's cluster membership to be determined.

Each remaining pattern is assigned to a cluster according to its maximum *Pattern-Cluster Membership* (PCM) value among all clusters. The PCM value for a pattern P with respect to a cluster C, denoted $PCM(P, C)$, is defined as the fraction of $PQ1(G(C))$ represented by P:

$$PCM(P, C) = \frac{|\text{mds}\,(P) \cap \text{mds}\,(G(C))|}{PQ1\,(G(C))}$$

(P's length ratio is unnecessary since it is a constant in P's PCM values for all clusters.) Conceptually, $PCM(P, C)$ measures the fraction of C's pattern-based description represented by P, or P's "prevalence" in C.

This step creates a complete CP set for each cluster C, denoted $PS(C)$. The union of these sets contains all frequent patterns except those whose maximum PCM values occur at two or more clusters. $G(C)$ remains unchanged in this step (i.e., $PS(C) \supset G(C)$), so patterns can be considered in any order in the computation of PCM values.

Step 4. Assign Tuples to Clusters: once frequent patterns are assigned to clusters as CPs, each tuple t of the dataset can be assigned to some cluster C. This is done based on each CP P's *vote* for t's membership in cluster C. The vote, denoted *vote*(P), reflects P's quality (measured by length ratio), P's prevalence in C (measured by PCM), and P's exclusivity to C. P's exclusivity to C is maximal if it belongs only to C, and minimal if it is nearly equally prevalent in another cluster. These qualities are captured in formula by:

$$vote(P) = \frac{PCM(P, C_{1st}) - PCM(P, C_{2nd})}{\sum_{i=1}^{k} PCM(P, C_i)} * \left(\frac{|P_{\max}|}{\mathsf{mgLen}(P)} \right)^2 .$$

Here, C_{1st} and C_{2nd} denote the two clusters respectively associated with P's highest and second-highest PCM values. By normalizing the difference of those two PCM values, we also take into consideration P's PCM values in other clusters.

Summing votes of all patterns for a single cluster C, we get t's *Tuple-Cluster Membership* (TCM) value for C:

$$TCM(t, C) = \sum_{P} \{vote(P) \mid P \in PS(C) \wedge t \in \mathsf{mds}(P)\} .$$

A tuple t is assigned to the cluster C that maximizes $TCM(t, C)$. [If t's highest TCM value is attained at multiple clusters, it can be assigned later by another method (e.g., a classification algorithm).]

We note that data preprocessing and frequent pattern mining are needed before using CPC. The "Frequent Equivalence Class Miner" [78] (based on FP-growth [92]) can be used for pattern mining.

8.6   GENERAL EXPERIMENTAL EVALUATION OF CPC

CPC was evaluated on the CPCQ scores of the computed clusterings as well as its ability to recover expert-given classes (measured by F-score). Reference [80] provided experimental results for two categorical and one numerical datasets, all from the UCI Repository. [The authors also considered Mushroom (categorical), Breast Cancer Wisconsin Diagnostic (numerical), and Statlog Heart (mixed-type), also from UCI, although the results on these were omitted.] Four other clustering algorithms, including Simple K-Means and EM, were used for comparison.

In summary, in all six datasets from UCI, CPC achieved the highest F-score, while in four of those six datasets, CPC achieved the highest CPCQ score. The experiments also showed that, in general, CPCQ which is an internal measure, can act as a surrogate of F-score, which is an external measure.

8.7   TEXT DATA ANALYSIS ON BLOGS USING CPC

Tables 8.1 and 8.2 illustrate CPC's ability to find good clusters and the usefulness of the cluster summaries produced by CPC for the clusters. These tables report results for text (weblog) clustering. The four categories of health, music, sports, and business, from the BlogCatalog [229] dataset were used. CPC performed clustering on the union of two or four categories (sets) of weblogs; the first row of Table 8.1 shows which two or four categories were used; and result in Table 8.2 is obtained on all four categories. Data was preprocessed by removing duplicate weblogs, removing stopwords, and stemming; words were treated as items.

Table 8.1 reports the F-scores as well as CPCQ values of the clustering results of CPC. Table 8.2 shows the first two CP groups G_1 and G_2 produced by CPCQ of Chapter 7 on the clustering computed by CPC.

Table 8.1: BlogCatalog: F-scores vs. CPCQ scores

$minSup =$	Health, Music		Sports, Business		Health, Music, Sports, Business	
	F-score	CPCQ	F-score	CPCQ	F-score	CPCQ
0.03	0.890	3.28	0.846	0.625	0.757	0.456
0.02	0.893	11.40	0.830	0.650	0.772	0.411
0.01	0.897	12.00	0.828	0.690	0.710	0.366

Table 8.2: Example cluster descriptions

	Cluster 1	Cluster 2	Cluster 3	Cluster 4
G_1	{busi, market}	{band, song}	{symptom}	{team, game}
G_2	{money, internet}	{love, song}	{people, disea}	{season, game}

Based only on these cluster descriptions, one can easily estimate what the clusters are mainly about: they are about business, music, health, and sports. The use of multi-item patterns such as {season, game} clearly helps with suggesting the themes of the clusters. (Each of the pattern groups contains just one multi-item pattern.) More details of these experiments can be found in [54].

8.8   DISCUSSION

This chapter presented a pattern-based clustering method called CPC. The algorithm starts with frequent patterns, it selects some frequent patterns to make them contrast patterns of certain clusters, and it does that to maximize the CPCQ value. Hence CPC forms clusters to maximize the richness of high-quality contrast patterns for the resulting clusters. Special strengths of the method include: (1) CPC does not use distance metrics, and, (2) the method produces interpretable results by associating each cluster with a set of contrast patterns characterizing the cluster. Interestingly, the method focuses its attention on the quality of the contrast patterns associated with clustering results, instead of distance-based quality metrics which are often used by traditional algorithms.

CHAPTER 9

IBIG: **Ranking Genes and Attributes for Complex Diseases and Complex Problems**

Gene ranking plays an important role in disease analysis[1] in general, and for identifying major contributing genes for given diseases in particular. Some diseases are hard to analyze since they are complex diseases, that is, they are mostly influenced by the interaction of multiple genes. Gene ranking for complex diseases is challenging due to the exponential number of potential interactions since the number of genes is in the thousands or more.

This chapter presents an emerging pattern based gene-ranking method called IBIG (Interaction-Based Importance of Genes), designed to perform gene ranking for complex diseases. IBIG ranks genes by taking multi-gene interactions into consideration, and a gene's rank given by IBIG is based on the number of influential interactions that the gene participates in. IBIG uses jumping emerging patterns (JEPs) to represent multi-gene interactions. Given a set of mined JEPs, a gene's IBIG rank depends on the number of high-quality JEPs involving the gene and the quality of those JEPs; a gene is ranked high if it occurs in many high-quality JEPs. The *quality* of a JEP is *measured by its support in its home class.*[2] In order to get the optimal ranking of genes, some optimal set of JEPs (that exist in the data underlying the disease) needs to be mined. IBIG does that by using an iterative algorithm that combines gene ranking and pattern mining using a so-called gene-club technique, which is especially designed to address the challenge posed by the presence of thousands of genes. We emphasize that the IBIG method often mines high-quality JEPs having much larger home-class support than other methods.

Organizationally, this chapter starts by defining the gene-ranking problem, followed by a brief review of representative gene-ranking methods. Then it gives some background on complex diseases. Next, it explains how emerging patterns can represent interactions among genes. Then it presents the IBIG gene-ranking method, together with the iterative algorithm and the gene-club technique. It then gives experimental findings obtained by IBIG on microarray data for

---

[1]Besides disease analysis, attribute ranking is also useful as an approach for feature selection. Hence it is useful for classification, recommendation, outlier detection, and so on.

[2]Recall that the support of a JEP in non-home classes is zero.

some cancer to illustrate the potential of IBIG. It also discusses how result of IBIG gene ranking differs from that of single-gene-based gene-ranking method. Finally, some concluding remarks are given.

The chapter's title is chosen to reflect the idea that the IBIG approach can also be used to rank attributes for complex problems[3] outside of medicine. The chapter mostly uses terminologies centered around genes to facilitate the presentation. The domain of each gene/attribute can be continuous or binary. For the discussion below, we typically discretize each numerical attribute into two intervals (which will be referred to as "low" and "high") using the entropy-based method [65].

While this chapter uses microarray gene-expression data in the discussion, the IBIG method can also work with other high dimensional biological/medical data that describes genes or other factors of interest. Such data can be obtained using different technologies and can be represented using different biological/medical features, including deep sequencing, copy number variations, single nucleotide polymorphisms (SNPs), and so on. See [157] for a survey on the various kinds of such data.

9.1   BASICS OF THE GENE-RANKING PROBLEM

We consider gene ranking by using data with classes. More specifically, we have a set $\mathcal{G} = \{g_1, \ldots, g_m\}$ of genes, two classes (positive and negative), a set POS of positive tuples, and a set NOR of normal tuples. Each tuple t is a vector of $|\mathcal{G}|$ numerical values, one for each $g \in \mathcal{G}$. The values for a gene can be continuous or binary (such as the case for SNP data). We use $t[g_j]$ to denote t's value for g_j. Sometimes it is convenient to combine the two classes into a dataset D of class-tagged tuples obtained by adding a class attribute whose value for a tuple is the class of the tuple.

A gene-ranking method produces a ranked list of the form $g_{i_1}, \ldots, g_{i_m}$ of the genes. This is often achieved by associating each gene with some computed numerical value and then sorting the genes in decreasing order of their associated values.

A gene ranked near the beginning of the ranked list is said to be ranked high; it is ranked low otherwise. We also use relative positions of genes in the ranked list to say things like "one gene is ranked lower than another gene." Sometimes we are only interested in the highly ranked genes up to a certain position.

We now describe two often used gene-ranking methods.

The fold-change method [79, 216] ranks genes based on their fold-change (FC) values defined by

$$FC(g) = \frac{avg(g, POS)}{avg(g, NOR)},$$

---

[3]We define a complex problem to be any problem where the underlying data is similar to complex diseases. As a result, solutions using interactions can be significantly better than solutions that do not use interactions.

where g is a gene, $avg(g, POS)$ is the average g value in POS, and $avg(g, NOR)$ is the average g value in NOR. For simplicity, we assume all gene values are positive.

The entropy-based method ranks genes based on their information gain (IG) values (see Chapter 2). For each gene g, $\mathsf{IG}(g)$ is defined to be

$$\mathsf{IG}(g) = \max_{v} \mathsf{IG}(g, v).$$

In the formula, v ranges over all split values for g, and

$$\mathsf{IG}(g, v) = \mathsf{entropy}(D) - \sum_{i=1}^{2} \frac{|D_i|}{|D|} \mathsf{entropy}(D_i),$$

where $D_1 = \{t \in D \mid t[g] \leq v\}$ and $D_2 = \{t \in D \mid t[g] > v\}$.

These two methods only use the relationship between a gene itself and the class to rank the genes. They do not consider interactions among genes, which are the key characteristics of complex diseases.

9.2 BACKGROUND ON COMPLEX DISEASES

The following quote, from a website at the National Institutes of Healthy (NIH),[4] defines complex diseases in an intuitive manner, lists some examples of complex diseases, and describes the current state of the understandings of such diseases.

> Nearly all conditions and diseases have a genetic component. ... Common medical problems such as heart disease, type 2 diabetes, and obesity do not have a single genetic cause—they are likely associated with the effects of multiple genes (polygenic) in combination with lifestyle and environmental factors. Conditions caused by multiple contributing factors are called complex or multifactorial disorders. ... Complex disorders are difficult to study and treat because the specific factors that cause most of these disorders have not yet been identified. Researchers continue to look for major contributing genes for many common complex disorders.

Besides those listed above, cancers and Parkinson's disease are also believed to be complex disorders.

As complex diseases are caused by a combination of multiple genes and other factors, it is harder to understand and analyze them and to identify major contributing genes than for Mendelian diseases that are each caused by one gene. Reference [147] notes that "increasing evidence supports the extensive and complex genetic contribution to Parkinson's disease (PD)," "the effect of each individual locus[5] is small," and "thus far only a small portion of the heritable

---

[4]https://ghr.nlm.nih.gov/primer/mutationsanddisorders/complexdisorders
[5]A locus refers to one particular SNP.

component of PD has been identified." It is likely that the above three statements are true for many, if not all, complex diseases. Reference [32] gives the following perspective on why complex diseases (in the name of complex traits) are hard to analyze:

> Many complex traits are driven by enormously large numbers of variants of small effects, potentially implicating most regulatory variants that are active in disease-relevant tissues. To explain these observations, we propose that disease risk is largely driven by genes with no direct relevance to disease and is propagated through regulatory networks to a much smaller number of core genes with direct effects.

Reference [157] gives a survey on various kinds of data and various tools that have been developed to understand and analyze complex diseases.

9.3   CAPTURING INTERACTIONS USING JUMPING EMERGING PATTERNS

Gene ranking for complex diseases should be useful for identifying their major contributing genes, and the rank of genes should reflect the role of genes in influential interactions for a given disease. By an interaction we mean a combination of several genes that strongly influence a given disease.

In statistics, an interaction[6] refers to a relationship among three or more variables, in which the simultaneous influence of two variables on a third is not additive. For data with classes, the third variable is often the class attribute. Interactions are often considered in the context of complex disease analysis (e.g., [172, 174, 200]) as well as in the context of regression analyses [97, 98].

We use a special kind of emerging patterns to represent interactions among genes. Specifically, we say an emerging pattern P represents an interaction (over the set of genes involved in P) if there is some class $C2$ (either $PPOS$ and NOR) such that growthRate($P, C2$) is large and it is much larger than growthRate($Q, C2$) for all proper sub-patterns Q of P. The requirement that growthRate($P, C2$) is large ensures that the pattern is a reliable marker for the disease. The requirement that "growthRate($P, C2$) is much larger than growthRate($Q, C2$) for all proper sub-patterns Q of P" ensures that the pattern represents a marker that cannot be replaced by other related patterns.

It is easy to see that each minimal jumping emerging pattern is an interaction according to the above definition. Indeed, a minimal jumping emerging pattern P has the infinite growth rate, and the condition that "growthRate($P, C2$) is much larger than growthRate($Q, C2$) for all proper sub-patterns Q of P" is satisfied since no proper subset of a minimal jumping emerging pattern P is also a jumping emerging pattern.

---

[6]https://en.wikipedia.org/wiki/Interaction_(statistics)

We note that emerging patterns were used to define interactions in [31], although that definition is somehow different from the one given above.[7]

9.4   THE IBIG APPROACH

9.4.1   HIGH-LEVEL VIEW OF THE IBIG APPROACH

To compute the optimal ranking of genes with emerging patterns representing interactions, naturally one would hope to use the highest quality emerging patterns that exist. However, there are serious challenges: there are often thousands of genes and no known algorithm can efficiently mine all emerging patterns when there are so many genes. The IBIG approach uses a set of high-quality emerging patterns, mined using a heuristic approach, to approximate the highest quality emerging patterns.

While the IBIG approach can use general emerging patterns, in the simplest setting, it can use just jumping emerging patterns (JEPs). The quality of a JEP is measured by its support in its home class (the class where its support is not zero).

The IBIG approach uses multiple iterations; in each iteration, it uses a gene-club based approach to mine JEPs and rank the genes using the currently available mined patterns. Since different genes are often involved in different gene-set interactions, in each iteration, the IBIG algorithm mines the JEPs over a number of different gene clubs. Each gene club is a small group of genes that are highly related to one gene (called the owner gene); it is computed for each of the currently highly ranked genes. A gene club is used to help find the highest support JEPs involving one owner gene, by performing JEP mining on the given dataset projected onto the set of genes in the gene club. Different gene clubs are used separately and independently. By performing JEP mining for all current highly ranked genes using their respective gene clubs, the IBIG approach can mine high-quality JEPs by working with the relatively large search space represented by the multiple gene clubs.

9.4.2   IBIG GENE RANKING BASED ON A SET OF EMERGING PATTERNS

Let $\mathcal{P}$ be a set of JEPs (typically a set mined in a given iteration of the IBIG algorithm). We now describe how to compute the IBIG ranking based on the set $\mathcal{P}$, which we denote by $IBIG_{\mathcal{P}}$. We note that the final IBIG ranking is computed by the iterative algorithm discussed in Section 9.4.4.

We first note that different JEPs may be highly similar to each other. This should be addressed in computing $IBIG_{\mathcal{P}}$. One way to address this issue is to compute IBIG ranking by using a highly diverse subset of $\mathcal{P}$. The IBIG ranking uses such a way; it selects that subset as follows. For each gene g it selects from $\mathcal{P}$ one random JEP having the highest support in its home class among the JEPs involving g. Let H denote the set of the selected JEPs after considering all genes.

---

[7]Several studies considered mining and analyzing interactions for complex diseases, including [73, 75, 173].

The *IBIG*$_\mathcal{P}$ value of a gene g is defined as

$$IBIG_\mathcal{P}(g) = \sum_{P \in H_g} supp^*(P),$$

where $H_g = \{Q \in H \mid Q$ involves $g\}$ and $supp^*(P) = supp(P, POS) + supp(P, NOR)$. Observe that only one of $supp(P, POS)$ and $supp(P, NOR)$ is not zero as P is a JEP.

The *IBIG*$_\mathcal{P}$ value for gene g is proportional to "how often g participates in high-quality interactions represented by JEPs" and the quality (home-class support) of those interactions represented by JEPs.

9.4.3   GENE CLUBS AND COMPUTING GENE CLUBS

A *gene club* of a gene g is a set of genes that are highly correlated with g; g is called the *owner* of the gene club, and the number of genes in the club is the *size* of the club. The owner gene g is always a member of g's gene club. Observe that, in general terminology, a gene club is a group of features or a feature group.

It should be noted that gene clubs for different genes of interest are selected separately and they are often very different from each other. Gene clubs for a given owner gene are often very different from the (global) sets of features selected by normal feature-selection methods.

There are many ways to compute gene clubs. Reference [137] considered four; we present the experimental winner, which is called the *iterative gene-club formation method* and denoted by GCIT.

The GCIT method forms a gene club by iteratively adding new genes having the largest information gain. Let $\{g_1, \ldots, g_m, g\}$ be a partial gene club for owner gene g. The *information gain* for a gene g' with respect to $\{g_1, \ldots, g_m, g\}$ is defined by

$$IG(g' \mid g_1, g_2, \ldots g_m, g) = IG(g_1, g_2, \ldots g_m, g, g') - IG(g_1, g_2, \ldots g_m, g).$$

The *IG* measure above is a generalization, which we define next, of the original *IG* (see Chapter 2). Let S be a subset of the given dataset D used for gene ranking. Let $g_1, g_2, \ldots g_n$ be a sequence of n genes. For each string $B = b_1 b_2 \ldots b_n$, where each b_i is either "low" or "high," of length n, let S_B be the set of tuples in S such that "$g_i = b_i$" is true for each i. Let $Entropy(S_B)$ denote the entropy of S_B. The *information* for gene group $\{g_1, g_2, \ldots, g_n\}$ is

$$I(g_1, g_2, \ldots, g_n) = \sum_{B} \frac{|S_B|}{|S|} Entropy(S_B),$$

where B ranges over all possible strings over $\{low, high\}$ of length n. The *information gain* for gene group $\{g_1, g_2, \ldots, g_n\}$ is

$$IG(g_1, g_2, \ldots, g_n) = Entropy(S) - I(g_1, g_2, \ldots, g_n).$$

The GCIT method finds a gene club of size k for g as follows: it first selects the gene g_1 having the highest $IG(g' \mid g)$ among all genes $g' \neq g$ and initializes the gene club to $\{g, g_1\}$; then, for $i = 2 \ldots k - 1$, it repeatedly adds the next gene g_i having the highest $IG(g' \mid g_1, \ldots, g_{i-1}, g)$ among all remaining genes g' to the current gene club.

9.4.4   THE ITERATIVE IBIG ALGORITHM: IBIGI

The pseudo-code of the iterative IBIG algorithm (IBIGi) is given in Algorithm 9.9. Step (1) of the algorithm discretizes each gene's range into two intervals (low and high), using entropy-based binning. *GK* of Step (2) can be user given or it can be the top-k genes computed by a gene-ranking method such as IG. In Step (3), emerging patterns are mined from the data projected on to *GK*. Step (4) computes the gene ranking using the current set of JEPs. Steps (5) and (6) compute the gene clubs for each of the top-k ranked genes computed in Step (4). In Step (7), for each of the just-computed gene clubs, emerging patterns are mined from the data projected onto the gene club. Step (8) forms the union of the JEPs mined from Step (7). The termination condition *TC* is "either the ranking does not change from the previous iteration or a certain number of iterations (e.g., 10) has been performed." If *TC* is not true, then the loop containing Steps (4), (5), (6), (7), and (8) is repeated.

Algorithm 9.9 IBIGi: The Iterative IBIG Algorithm

Parameters: k (top k genes) and ℓ (gene club size)
Input: Dataset D with classes
Initialize:
    (1) Discretize the genes
    (2) Select a set *GK* of k genes
    (3) Mine the emerging patterns *EPS* involving genes in *GK*
Repeat
    (4) Perform IBIG gene ranking using *EPS*
    (5) For each of the top-k genes g_i (of the current IBIG ranking)
        (6) Compute g_i's gene club GC_i of size ℓ
        (7) Let EPS_i be the mined EPs involving genes in GC_i
    (8) Let EPS be the union of $EPS_1, \ldots, EPS_k$
Until the termination condition TC is satisfied

In experiments discussed below, both k and ℓ were set to 20. For Steps (3) and (7), the BorderDiff algorithm [56] was used to mine the jumping emerging patterns. For Step (2), *GK* is given by IG.

9.5 EXPERIMENTAL FINDINGS ON IBIG ON COLON CANCER DATA

We now report four main experimental findings to indicate the need for, and the potential of, the IBIG gene-ranking approach and the IBIGi algorithm. This is done using the colon cancer dataset [14], which contains 2,000 genes. Two findings (in Section 9.5.1) are concerned with the distribution of genes in high-quality JEPs (representing multi-gene interactions) and the effectiveness of the IBIGi algorithm, and the other two findings (in Section 9.5.2) are on the significant difference between IG-based gene ranking and IBIG gene ranking.

Recall that we use the home-class support of JEPs to indicate their quality.

9.5.1 HIGH-QUALITY JEPS OFTEN INVOLVE LOWLY IG-RANKED GENES AND IBIGi CAN FIND MANY OF THEM

As discussed in [119], when we limit ourselves to the top 35 IG-ranked genes, the highest support of the JEPs of the normal class is 77%, and that of the cancer class is 70%. The quality of the JEPs is lower if $k = 20$ is used. We note that, for 20 or 35 genes, we can easily mine all possible JEPs.

In contrast, if we go beyond the top 35 IG-ranked genes, the IBIGi algorithm can find JEPs of 100% support (see Tables 9.1 and 9.2) for both the normal class and the positive class. Moreover, such JEPs can be found by using the IBIGi algorithm with gene club size $\ell = 20$ and considering only gene clubs of the top $k = 20$ genes (starting with the IG rank). We also conducted an experiment to see how many high-quality JEPs that involve just the top 75 IG-ranked genes (which can be found easily) can be found by the IBIGi algorithm when using the limited $k = 20$ and $\ell = 20$ setting; the result indicates that the IBIGi algorithm can find many of those high-quality JEPs.

Finally, each of those mined JEPs with 100% support involves at least one gene whose IG rank is 113 or larger, and such mined JEPs involve genes whose IG rank is as large as 700. The largest IG rank of genes involved in the best 30 mined JEPs of the positive class mentioned above is $1,089$, and that of the normal tissue class is $1,261$. See Tables 9.1 and 9.2 for details.

To summarize, we have these two findings:

1. High-quality JEPs often involve lowly IG-ranked genes, and the JEPs limited to the top-k IG-ranked genes may have much lower quality than the best JEPs that exist.

2. The IBIGi algorithm can mine many JEPs of very high quality, much higher than the quality of JEPs limited to the top-k IG-ranked genes.

The importance of these findings rest on the assumption that the high-quality JEPs represent multi-gene interactions.

Table 9.1: JEPs with highest support—colon diseased (positive) class. Support in the normal class is 0 for all listed JEPs. Each integer in the patterns is the IG rank number of a gene. +: the gene's value is in the high interval; −: the gene's value is in the low interval.

JEP	Support in Positive Class	JEP	Support in Positive Class
{1+ 4- 112+ 113+}	100	{1+ 4- 113+ 116+}	100
{1+ 4- 113+ 221+}	100	{1+ 4- 113+ 696+}	100
{1+ 108- 112+ 113+}	100	{1+ 108- 113+ 116+}	100
{4- 108- 112+ 113+}	100	{4- 109+ 113+ 700+}	100
{4- 110+ 112+ 113+}	100	{4- 112+ 113+ 700+}	100
{4- 113+ 117+ 700+}	100	{1+ 6+ 8- 700+}	97.5
{1+ 8- 110+ 112+}	97.5	{1+ 8- 112+ 216+}	97.5
{1+ 8- 112+ 222+}	97.5	{1+ 8- 112+ 700+}	97.5
{1+ 8- 112+ 1089-}	97.5	{1+ 8- 116+ 1089-}	97.5
{1+ 110+ 116+ 263-}	97.5	{1+ 112+ 113+ 263-}	97.5
{1+ 112+ 263- 1089-}	97.5	{1+ 116+ 263- 1089-}	97.5
{4- 8- 112+ 216+}	97.5	{4- 112+ 113+ 263-}	97.5
{4- 112+ 113+ 1089-}	97.5	{6+ 8- 113+ 116+}	97.5
{6+ 8- 113+ 696+}	97.5	{6+ 8- 113+ 700+}	97.5
{8- 38+ 112+ 216+}	97.5	{8- 113+ 114+ 222+ 700+}	97.5

9.5.2 SIGNIFICANT GENE-RANK DIFFERENCES BETWEEN IG AND IBIG

Table 9.3 compares the IBIG rank against the IG rank for the colon data. We observe that the gene ranked as the 10th most important by IBIG was ranked at 401 by IG. The genes ranked at 2 and 3 by IBIG are ranked at 114 and 115 by IG.

Table 9.3 indicates the following:

1. Some individually lowly ranked genes are ranked very high when we consider interactions among genes. In other words, some genes may not be strongly related with the disease classes individually, but they are very important when combined with other genes.

2. Many individually highly ranked genes are ranked very low when we consider interactions among genes.

Table 9.2: JEPs with highest support—colon normal class. Support in the positive class is 0 for all listed JEPs. Each integer in the patterns is the IG rank number of a gene. +: the gene's value is in the high interval; −: the gene's value is in the low interval.

JEP	Support in Normal Class	JEP	Support in Normal Class
{12- 21- 35+ 40+ 137+ 254+}	100	{12- 35+ 40+ 71- 137+ 254+}	100
{20- 21- 35+ 137+ 254+}	100	{20- 35+ 71- 137+ 254+}	100
{5- 35+ 137+ 177+}	95.5	{5- 35+ 137+ 254+}	95.5
{5- 35+ 137+ 419-}	95.5	{5- 137+ 177+ 309+}	95.5
{5- 137+ 254+ 309+}	95.5	{7- 21- 33+ 35+ 69+}	95.5
{7- 21- 33+ 69+ 309+}	95.5	{7- 21- 33+ 69+ 1261+}	95.5
{7- 34- 35+ 69+}	95.5	{7- 34- 69+ 309+}	95.5
{12- 34- 35+ 69+ 136-}	95.5	{12- 35+ 40+ 188- 254+}	95.5
{12- 35+ 69+ 136- 309+}	95.5	{18- 33+ 35+ 40+ 254+}	95.5
{18- 33+ 254+ 309+}	95.5	{21- 35+ 188- 254+}	95.5

Table 9.3: IBIG rank vs. IG rank for colon cancer data. Gene No.: the position number of the gene in the original data.

IBIG Rank	Gene No.	IG Rank	IBIG Rank	Gene No.	IG Rank
1	1422	8	11	1634	34
2	681	114	12	764	4
3	575	115	13	1559	132
4	1670	1	14	257	12
5	1041	7	15	1885	212
6	1923	242	16	492	3
7	1581	16	17	248	2
8	624	6	18	580	39
9	1632	217	19	1327	66
10	174	401	20	398	14

9.6   DISCUSSION

From a methodological perspective, this chapter presented a gene-ranking method called IBIG to rank genes (and attributes) for complex diseases (and complex problems). The IBIG method reflects the importance of genes by considering their participation in influential multi-gene interactions. The IBIG method is emerging pattern based: it uses jumping emerging patterns (JEPs) to represent interactions. The associated IBIGi algorithm does two things: mining high-quality JEPs using gene clubs (which are groups of features strongly correlated with an owner gene) and using the mined JEPs to perform gene ranking. The IBIGi algorithm is iterative.

As feature/gene ranking can be used as a feature-selection method, this chapter also contributes to feature selection for complex diseases and complex problems.

From an application perspective, the chapter also discussed some notable findings that have potential in medical research and general problem solving. It was noted that, for complex diseases, (1) lowly ranked genes according to the information gain rank (IG) may be ranked very high according to IBIG (since they participate in many interactions represented by JEPs), and, (2) highly ranked genes according to the information gain rank (IG) may be ranked quite low according to IBIG. Genes satisfying (1) above can be good candidates for analyzing complex diseases and for identifying major contributing genes. The high-quality JEPs mined by IBIGi can also help us to identify multi-factor risk factors.

The IBIGi algorithm and the ideas for IBIG ranking were introduced in [137]. The IBIG ranking approach was presented in [138], although the focus of that article was on mining high-quality emerging patterns in the presence of thousands of genes.

CHAPTER 10

CPXR and CPXC: Pattern Aided Prediction Modeling and Prediction Model Analysis

This chapter presents pattern aided prediction (PXP) models, and the CPXP approach, which uses contrast patterns to construct PXP models. Here, prediction covers both regression and classification. Moreover, patterns are used as conditions, to define and characterize subpopulations whose local prediction models are substantially different (a) from local models for other subpopulations, and, (b) from the global model on the whole population (covering all data). Different from traditional boosting, the CPXP approach uses opportunity and reward to guide the boosting process. Different from traditional ensembles, the PXP models are conditional ensembles, as each member model is only applied to instances matching its associated pattern. The chapter also presents a so-called diverse predictor-response relationship concept. The concept of subpopulationwise conditional correlation analysis, involving a special adaptation of PXP and CPXP, is also discussed. Experimental results are summarized and discussed, including practical applications to water content prediction for soil and to medical risk prediction for both traumatic brain injury and heart failure.

At a high level, a pattern aided prediction model is roughly a small set of pairs of patterns and local models. The diverse predictor-response relationship concept characterizes prediction problems whose underlying datasets are each highly heterogeneous with respect to the given prediction problem. The fact that PXP models have good performance on many prediction problems shows that many prediction problems contain diverse predictor-response relationships.

Organizationally, the chapter first presents some background material. It then presents the concepts of PXP models, and the concepts of diverse predictor-response relationships. It then presents the CPXP algorithm. Experimental results are then discussed, followed by discussion on how to solve the subpopulationwise conditional correlation problem. Finally, some concluding remarks are given.

For comparison, we note that in Chapter 4 contrast/emerging patterns were used as simple classifiers, whereas in this chapter they are used as subpopulation "handles" for characterizing and labeling subpopulations and for representing boosting opportunities.

10.1   BACKGROUND MATERIALS

In this section we focus our attention to regression; information on classification can be found in Chapter 4.

Regression and classification are both supervised learning problems. The training dataset is a finite set of some s pairs of the form $\{(x_i, y_i) \mid 1 \leq i \leq s\}$. The x_i's are values for the predictor or input variables. For regression the y_i's are numerical values for the response variable. For classification the y_i's take classes as values. For both problems the aim is to learn a model to predict y from x. For classification the model is often called a classifier or classification model, whereas for regression the model is called a regression model.

Regression and classification use different measures to evaluate model performance. Regression often uses RMSE (root mean squared error): given a regression model f and a dataset $D = \{(x_i, y_i) \mid 1 \leq i \leq m\}$ for model evaluation, $RMSE(f)$ is defined to be $sqrt(\frac{\sum_{i=1}^{m}(f(x_i)-y_i)^2}{m})$. For any (x, y) pair, $f(x) - y$ is a residual (prediction error). Classification can be measured in many ways, including by accuracy or AUC (area under the curve).

Representative model types and methods for regression include Multiple Linear Regression (LR), Piecewise Linear Regression (PLR), Bayesian Additive Regression Trees (BART), Gradient Boosting (GBM), and Support Vector Regression (SVR). References for these can be found in [61]. A main direction of recent development has been regularization approaches, which use constraints to limit the total magnitude of the coefficients for all variables.

10.2   PATTERN AIDED PREDICTION MODELS

10.2.1   FITTING LOCAL MODELS FOR LOGICAL SUBPOPULATIONS

An important idea for "pattern aided prediction models" is to use several patterns to represent several subpopulations of data in order to capture highly different behavior using a respective local model for each subpopulation.

The subpopulations should be logical to ensure that "pattern aided prediction models" are both useful and interpretable, to help improve computational efficiency, and to exclude arbitrary, especially contrived, clusters as subpopulations.

Logical subpopulations can be defined and characterized using (frequent) patterns. For each pattern P the corresponding subpopulation is mds(P).

Each logical subpopulation, and its corresponding defining pattern, represents an opportunity, where one may be able to get a model to yield improved prediction accuracy compared to an original given model. Such a model is especially built for, and to be used on, a subpopulation. Hence such a model will be called a *local model*.

For each pattern P, we use h_P to denote a *local model* that is built on the corresponding subpopulation mds(P), using some selected regression or classification algorithm. This local model will only be applied to instances satisfying P.

The term "local" is used to mean that h_P will not be applied to instances not satisfying P. It has nothing to do with distance.

In experiments it was found that searching over contrast patterns linked to "large prediction error" is effective and efficient; it is more efficient than searching over frequent patterns. It should be more efficient than searching over arbitrary clusters, as the number of arbitrary clusters is much larger than the number of contrast patterns.

10.2.2 PATTERN AIDED PREDICTION MODELS

Intuitively, a pattern aided prediction model is a *cooperative* set of local models each operating on a subpopulation defined by a pattern. Different subpopulations can have overlap, and weighted voting is used to combine the (possibly multiple) predictions made by the local models. The two definitions below define the form and "semantics" of pattern aided prediction models, respectively.

Definition 10.1   A *pattern aided prediction model* (PXP) is a tuple

$$M = ((P_1, h_{P_1}, w_1), \ldots, (P_k, h_{P_k}, w_k), h_d),$$

where $k > 0$ is an integer, $P_1, \ldots, P_k$ are patterns defining k subpopulations, h_d is the *default model*, and for each i, h_{P_i} is P_i's *local model* and $w_i > 0$ is P_i's *weight*.

Definition 10.2   Consider M given in Definition 10.1. For each data instance x, let

$$\pi_x = \{P_i \mid 1 \leq i \leq k, x \text{ matches } P_i\}.$$

If $\pi_x = \emptyset$, then M uses h_d to perform prediction on x. Otherwise ($\pi_x \neq \emptyset$), M uses weighted voting of the local models associated with patterns in π_x to perform prediction on x. More specifically:

(a) If the underlying problem is regression, then

$$weighted\_vote_s(M, x) = \frac{\sum_{P_i \in \pi_x} w_i * h_{P_i}(x)}{\sum_{P_i \in \pi_x} w_i}.$$

(b) If the underlying problem is classification and $C_1, \ldots, C_\kappa$ are the classes, then, for each C_j,

$$weighted\_vote_s(M, C_j, x) = \frac{\sum_{P_i \in \pi_x} w_i * h_{P_i, C_j}(x)}{\sum_{P_i \in \pi_x} w_i}.$$

For classification, $h_{P_i, C_j}(x)$ denotes a predicted probability for "x belongs to C_j."

Example 10.3   Consider a regression problem with u, v, z being the predictor variables and y being the response variable. Let $w_1 = 0.6$, $w_2 = 0.3$,

$$P_1 : 2 \leq u < 5 \,\&\, 3 \leq v < 7;$$
$$P_2 : 6 \leq z < 9;$$
$$f_1 : y = 4 + 3z;$$
$$f_2 : y = 5 + 2u;$$
$$f_d : y = 2u + 4v.$$

Then $M = ((P_1, f_1, w_1), (P_2, f_2, w_2), f_d)$ is a PXP model (for regression). For $x = (u, v, z)$, $f_{PM}(x) = f_1(x)$ if x satisfies P_1 only, $f_M(x) = f_2(x)$ if x satisfies P_2 only, $f_M(x) = (0.6f_1(x) + 0.3f_2(x))/0.9$ if x satisfies both P_1, P_2, and $f_M(x) = f_d(x)$ otherwise (x satisfies none of P_1 and $_P2$).

In [61] PXP was called PXR (for regression), and in [62] PXP was called PXC (for classification).

A PXP is conditional ensemble in the following sense: each member h_P is only applied on an instance x if P matches x. This is different from traditional ensembles where each ensemble member is applied to all instances. Such conditional ensembles are powerful, and they are more flexible than traditional ensembles. They are often much smaller (often using just around seven patterns in experiments).

The patterns in a PXP do not need to use the whole set of variables, and different patterns do not need to use the same set of variables. They often just need to involve three or four variables, and those variables are selected automatically depending on the unique properties of the data.

The way PXPs are defined makes them especially suited to prediction problems having diverse predictor-response relationships (see Section 10.5).

10.3   CPXP: CONTRAST PATTERN AIDED PREDICTION

This section presents the CPXP algorithm. It should be noted that, in [61] CPXP was called CPXR (for regression), and in [62] CPXP was called CPXC (for classification).

The main idea of CPXP is to look for patterns P (representing focused *opportunities*) such that the baseline model makes lots of large prediction errors on instances in $\mathsf{mds}(P)$ and some local models can be found to correct those errors significantly (representing *rewards*), and to select a cooperating subset of the discovered patterns (and local models) to make a PXP model. The baseline model is just an original model built on the whole dataset (or given by the user).

We now give brief explanations of Algorithm 10.10. CPXP takes a training dataset D as input, and it takes two parameters: a ratio $\rho > 0$ for dividing D into large error (LE) and small error (SE) parts, and a *minSup* threshold on support of contrast patterns in *LE*. Step (1) builds a baseline model using an algorithm specified by the user. Step (2) divides D into *LE*, *SE* based on h_b's errors, so that the cumulative error of instances in *LE* divided by the cumulative error of all instances is $\geq \rho$. (Clearly instances in *LE* should have larger errors than instances in *SE*.) Step (3) performs binning and contrast pattern mining. Entropy-based binning is preferred. The set of contrast patterns whose support in *LE* is $\geq minSup$ and whose *growthRate* is ≥ 1 is taken as the initial set of contrast patterns. Step (4) reduces the set of mined contrast patterns, mostly based on mds overlap. Step (5) learns a local model for the mds of each remaining pattern. To avoid overfitting, the algorithm does not use a pattern P if the cardinality of mds(P) is smaller than the number of variables/attributes. Step (6) removes contrast patterns that offer little error reduction. Step (7) selects an optimal pattern set $S = \{p_1, \ldots, p_k\}$ to define a PXP. Step (8) determines pattern weights; each w_i can be defined as the relative error reduction on instances in mds(P_i) by h_{P_i} vs. the baseline model. Step (9) learns a default model h_d on instances not matching any P_i. Step (10) returns the computed PXP model.

Algorithm 10.10 The CPXP Algorithm

Parameters: ρ (for dividing D into LE and SE), *minSup* (for CP mining)
Input: D (training dataset)
Output: A PXP model
   (1) Learn a baseline model h_b on D
   (2) Divide D into *LE*, *SE* based on h_b's errors
   (3) Perform binning and contrast pattern mining
   (4) Reduce the set of mined contrast patterns
   (5) Learn local models for remaining patterns
   (6) Remove contrast patterns of low utility
   (7) Select optimal pattern set $S = \{P_1, \ldots, P_k\}$
   (8) Determine pattern weights, $w_1, \ldots, w_k$
   (9) Learn a default model h_d
   (10) Return the PXP model $((P_1, h_{P_1}, w_1), \ldots, (P_k, h_{P_k}, w_k), h_d)$

    The pattern set search part is expensive and many methods can be used. Moreover, different objective functions for the search can be used. CPXP uses a heuristic iterative process to search for a pattern set $S = \{P_1, \ldots, P_k\}$ to construct a PXP model. CPXP uses average prediction error reduction as the objective function. Roughly speaking, the iterative search process involves two nested loops. Each iteration of the outer loop selects one pattern P that maximizes the objective function to add to the current pattern set S. After each addition, CPXP uses an inner loop to repeatedly select a pattern pair Q, P_Q, where $Q \in S$ and $P_Q \in CPS - S$, such that replacing Q by P_Q gives the largest increase to the objective function. The inner loop terminates if its last iteration did not improve the objective value by more than 0.1%, and the outer loop terminates if its last iteration did not improve the objective value by more than 1%.

The search process of CPXP automatically determines the number of patterns in the constructed PXP models—it is the number of patterns in S when the search process converges. The number of patterns can also be given as a parameter to CPXP.

CPXP will be affected by decisions on the following issues: the way to measure prediction (especially classification) error, the algorithms to be used for building baseline/local models, and the objective functions to be used in the pattern set search process. The details can be found in [61, 62]. We note that multiple linear regression, piecewise regression, and logistic regression algorithms can all be used for building baseline/local regression models, and logistic regression, decision trees, and Naive Bayes can all be used for building baseline/local classification models.

CPXP can take a prediction model h_b from the user, in which case CPXP is used to analyze the given prediction model h_b. Thus, it can be used for prediction model analysis.

10.4   RELATIONSHIP WITH BOOSTING AND ENSEMBLE MEMBER SELECTION

Recall that CPXP's main idea is to look for patterns P (representing focused *opportunities*) such that the baseline model makes lots of large prediction errors on $\mathsf{mds}(P)$ and some local models can be found to correct those errors significantly (representing *rewards*). So CPXP boosts the performance of an original prediction model by examining a well focused set of opportunities and using rewards to perform selecting which opportunities to use. This is *opportunity-focused and reward-guided boosting*. Moreover, CPXP selects a *cooperating set* of patterns (opportunities) based on their mds overlap and on improvement on prediction accuracy to make the final conditional ensemble.

10.5   DIVERSE PREDICTOR-RESPONSE RELATIONSHIPS

To consistently achieve high accuracy, a major challenge (to many regression algorithms and classification algorithms) is that many prediction problems contain "diverse predictor-response relationships":

> A prediction problem is said to have *diverse predictor-response relationships* (DPR relationships) if some optimal local prediction models fitting distinct logical subpopulations (of substantial sizes) of the underlying data are highly different from the optimal global prediction model for the problem. Some of the optimal local prediction models for distinct logical subpopulations can also be highly different from each other.

We observe that four restrictive conditions are included in the above informal definition. (1) The subpopulations must be logical. This restriction is imposed to make the local models of the subpopulations more useful and interpretable, and to exclude arbitrary contrived clusters as subpopulations. (2) The subpopulations must be large enough according to some size threshold.

(3) The subpopulations must be highly different, which means that their mds' are fairly disjoint. Their local models can also be highly different. (4) The local models for different subpopulations must be very different from the global model.

Model difference should be primarily evaluated from an accuracy perspective: suppose for each $i \in \{1, 2\}$ we found a logical subpopulation defined by P_i and a corresponding optimal local model f_i fitted on $\mathsf{mds}(P_i)$. Then f_1 should have a much lower prediction accuracy on $\mathsf{mds}(P_2)$ than f_2 and vice versa. Moreover, differences on subpopulations and local models should also be evaluated in terms of differences on variables and coefficients, in the sense that the patterns use very different sets of variables, the local models use fairly different sets of variables, and the local models use very different coefficients on shared variables.

The diverse predictor-response relationships concept is related to the ensemble diversity concept [104] from the ensembles community. But there is significant difference since the former refers to local models on logical subpopulations, whereas the latter is concerned with ensemble members that are "universal models" (which are meant to be applied to all instances for the entire underlying prediction problem). The diverse predictor-response relationship concept is also linked to the concept of "heterogeneous diseases" [15, 108] in the field of medicine.

The DPR relationship concept was first presented in [61]. The above version is more polished. Reference [61] used 25+% relative prediction error reduction by CPXR over LR to indicate that the underlying problem contains DPR relationships.

10.6   USES OF CPXR AND CPXC IN EXPERIMENTS

Below we discuss two kinds of experiments involving CPXR and CPXC. The first kind involves datasets previously used in regression and classification model evaluations. The second type involves practical utilization of CPXP to solve challenging prediction and analysis problems in agriculture science and in medicine/healthcare.

10.6.1   EXPERIMENTS ON COMMONLY USED DATASETS

Reference [61] reported experiments involving 50 real datasets obtained from several sources; 43 were from [100], and several were from the UCI repository. These experiments mainly compared performance of various regression algorithms on RMSE reduction defined by $\frac{RMSE(LR)-RMSE(M)}{RMSE(LR)}$, where $RMSE(M)$ is the RMSE of the model built by a given algorithm called M and LR refers to the standard multiple linear regression algorithm. The experiments found that CPXR (CPXP for regression) achieved the highest average RMSE reduction of 42.89%, whereas the best competing method achieved 20.18%. Moreover, out of the 50 datasets, there are 42 where CPXR's RMSE reduction is over 25%. This means that the RMSE reduction of the pattern aided models is 25% better than using one LR model. This indicates that many datasets for regression contain diverse predictor-response relationships. Moreover, CPXR also was found to outperform other methods on noise resistance and hence also on overfitting avoidance.

Reference [62] performed experiments on a range of classification datasets from UCI and it found that CPXC (CPXP for classification) also had much better performance than traditional classification algorithms.

PXP models often contain between 7 and 10 pattern-model pairs; as conditional ensembles they contain a very small number of members, making them easy to explain and understand. This is a big advantage compared with traditional ensembles, which often contain hundreds of ensemble members.

10.6.2 APPLICATIONS FOR AGRICULTURE AND HEALTHCARE PREDICTIONS

Reference [194] used CPXR (using logistic regression) to develop prognostic risk models to predict 1-, 2-, and 5-year survival of heart failure (HF) patients using data from electronic health records (EHRs). It found that one of the models generated by CPXR achieved an AUC and accuracy of 94% and 91%, respectively, which are significantly better than prognostic models reported in prior studies. Data extracted from EHRs allowed the incorporation of patient co-morbidities into the PXR models. The risk models developed by CPXR also reveal that HF is a highly heterogeneous disease, in other words, there exist different subgroups of HF patients who require different types of considerations with their diagnosis and treatment. That is, the disease contains diverse predictor-response relationships. CPXR can identify and properly handle such relationships. The study [194] led to two valuable insights for application of predictive modeling techniques in biomedicine: logistic risk models often make systematic prediction errors, and it is prudent to use subpopulation-based prediction models such as those given by CPXR when investigating heterogeneous diseases.

Reference [193] applied CPXR (using logistic regression) for predicting response to planned treatment on patients of traumatic brain injury, and it also found that CPXR has very good performance.

Reference [86] applied CPXR to build regression models (called pedotransfer functions in the paper) for soil water retention curve (SWRC) and other "water prediction problems for soil." In particular, the paper wanted to check if some easy-to-measure variables such as sample dimensions are useful predictor variables. (Some predictor variables are expensive to measure.) The study found that, including variables on sample dimensions (such as sample internal diameter and height) substantially improved the accuracy of the prediction accuracy of the models developed using the CPXR method. Moreover, the CPXR models including variables on sample dimensions have much better performance than the multiple linear regression models using identical sets of input variables. Indeed, Table 5 of the cited paper indicates that the RMSE of two PXR models are 0.034 and 0.027, respectively, whereas the RMSE of two multiple linear regression models are 0.061 and 0.060, respectively. Again the results indicate that these prediction problems contain diverse predictor-response relationships and CPXR can identify and properly handle such relationships.

10.7 SUBPOPULATIONWISE CONDITIONAL CORRELATION ANALYSIS

Reference [52] introduced the concept of subpopulationwise conditional correlation analysis. It argued that traditional correlation analysis is overly simplistic—it limits itself to examining correlation relationships over the entire underlying data space all together. The concept of subpopulation-wise correlations (SCRs) allows one to capture unusual correlations in subpopulations. The paper also adapted CPXR to mine such correlations.

Table 10.1 illustrates subpopulation-wise correlations between the prices of Gold and the Standard & Poor 500 index for the period between 2007 and 2017. The patterns also involve other financial variables.[1] The table lists five subpopulations, each defined by a pattern; the correlation for each subpopulation is represented by a linear regression model; the Pearson correlation coefficients (denoted by CC) are also listed. In the entire period, the correlation coefficient for *Gold* and *SPY* (given in the last row of the table) is 0.0896, which is only about 10% of the value

Table 10.1: Subpopulation-wise correlation between Gold and SPY [52]

Subpopulation ID	Pattern P_i, Correlation Models M_i		CC
S1	P_1:	$103.2293 \leq SPY < 148.8261$	0.3954
	M_1:	$Gold = 2.2475 + 1.1337 * \text{SPY}$	
S2	P_2:	$31.1272 \leq EWJ < 38.0454$	0.8711
	M_2:	$Gold = -27.4408 + 1.5112 * \text{SPY}$	
S3	P_3:	$1.1211 \leq Trea10Yr < 2.2422$	-0.9044
		$\& \ 0.0000 \leq Trea2Yr < 0.9859$	
		$\& \ 5.7842 \leq NatGas < 131.4251$	
	M_3:	$Gold = 226.2973 - 0.5373 * \text{SPY}$	
S4	P_4:	$79.2249 \leq Dollar < 87.3010$	0.3415
	M_4:	$Gold = 94.7121 + 0.2616 * \text{SPY}$	
S5	P_5:	$0.0000 \leq Trea2Yr < 0.9859$	0.2360
		$\& \ 52.3947 \leq VTI < 76.1176$	
		$\& \ 5.7842 \leq NatGas < 131.4251$	
	M_5:	$Gold = 121.2390 + 0.2758 * \text{SPY}$	
Default	P_d:	$not(P1 \lor P2 \lor P3 \lor P4 \lor P5)$	0.8402
	M_d:	$Gold = 69.0615 + 0.2024 * \text{SPY}$	
Global	M_g:	$Gold = 115.0426 + 0.0490 * \text{SPY}$	0.0896

[1]Data sources: https://www.investing.com/, and https://www.treasury.gov/.

of 0.8711 in subpopulation S2. The correlation coefficients for S2 and S3 have large comparable magnitudes but they have opposite signs (so one is a strong positive correlation and the other is a strong negative one). The default row represents the subpopulation of instances that are not in any of the five listed subpopulations.

10.8   DISCUSSION

This chapter presented the pattern aided prediction (PXP) models and a contrast pattern based mining algorithm (CPXP) to build PXP models. Patterns were used as conditions, to define and characterize subpopulations whose local prediction models are substantially different from the global model on all data, and whose local models for different subpopulations may also be very different. It presented the so-called diverse predictor-response relationship concept, to characterize prediction problems whose underlying datasets are each highly heterogeneous. Experimental results were discussed, including practical applications to water content prediction for soil and to medical risk prediction for both traumatic brain injury and heart failure.

The strong performance of PXP models constructed by CPXP can be attributed to their structure, which is highly suited to represent diverse predictor-response relationships, and to the opportunity-reward guided way of construction by CPXP.

The pattern-aided subpopulation-focused approaches presented in this chapter have potential for modeling and computing other kinds of knowledge types concerning subpopulations. This is especially true in the era of big data, due to the massive volume of, and the huge number of, attributes/variables contained in big data. The high dimensionality of, together with the complexities and heterogeneities of the complex structures embedded in, big data, make it very challenging regarding how to model such complex structures and how to extract them.

CHAPTER 11

Other Approaches and Applications Using Emerging Patterns

There are many interesting approaches and applications that involve the use of emerging/contrast patterns, and they are often quite different from each other. This chapter gives an overview of representatives of these approaches and applications; the presentation contains some details, but it is not very detailed due to space limitations. From a technique-focused perspective, the sections belong to six groups: (1) Sections 11.1–11.7 discuss representative approaches that use emerging/contrast patterns as group signatures/characterizations. They use such patterns to analyze given data groups, and perhaps also to solve other domain-specific problems. (2) Sections 11.8–11.14 are concerned with representative applications that use emerging patterns as features. (3) Sections 11.15–11.17 are about using CAEP to solve a range of classification/recognition problems. (4) Section 11.18 is about various ways to use emerging patterns to build classifiers. (5) Section 11.19 is about using emerging patterns for classification over streaming data. (6) Section 11.20 is about other applications. Section 11.21 provides a high-level view, from a "discipline and direction" focused perspective.

Materials already covered earlier may be included here, but not always, to give the big picture of the applications.

To make it easier for readers to locate applications/techniques of interest, this chapter is presented using quite a large number of sections, often using distinctive application-focused section titles. There is no straightforward way to organize the sections into groups, since they often involve different problems, different techniques, and different applications.

As it is focused on techniques/applications that use emerging/contrast patterns, this book makes no attempt at a survey of the mining algorithms of such patterns.

11.1 COMPOUND ACTIVITY ANALYSIS

Emerging patterns and the CAEP methodology were adapted in chemoinformatics as emerging chemical patterns (ECP) for the classification of active compounds [18]. It was found that the CAEP approach yielded accurate prediction models on the basis of much smaller training sets of active compounds than other machine learning methods [18] (also see Chapter 5), which made the CAEP approach attractive for classification tasks when only little information was available/required for model building. Accordingly, the CAEP/ECP approach was subsequently applied in iterative screening experiments for the identification of drug candidates [20] and in analyzing bioactive compound conformations [19]. More recently, the methodology was also used to classify compounds with multi-target activities [149], to predict individual compounds forming activity cliffs [150], and to classify compounds in different local structure-activity relationship environments [148].

11.2   STRUCTURE-ACTIVITY RELATIONSHIP EXPLORATION AND ANALYSIS

The analysis of structure-activity relationships (SAR) is one of the most important tasks during the early stages of the drug discovery process [140]. Pharmacophore modeling (in medicinal chemistry) is now a very well accepted approach for such investigations. Given a specific target, ligand-based pharmacophore elucidation requires the detection of the spatial arrangement of a combination of chemical features that are shared by several active molecules and that are responsible for favorable interactions with an active site of interest. Since pharmacophores are intrinsically three-dimensional, a key step of pharmacophore modeling is the conformational sampling of the molecules to be used for the alignment and the detection of the common features. Numerous approaches have been developed, but it is well recognized that broad conformational coverage requires an intensive computation when dealing with a large number of molecules. To overcome this limitation, Reference [140] decided to approach the pharmacophore elucidation problem by considering 2D molecular structures. Given a data set of molecules, the proposed method first enumerates all the recurring combinations of chemical features, then it computes their capability to discriminate between active and inactive molecules. Such recurring combinations of chemical features are captured using emerging patterns, and the discriminative capability is computed using the growth rates of emerging patterns. This work introduced an intuitive and interpretable visualization tool to analyze the SAR information, together with a pattern-based classifier relying on a selection of representative pharmacophores to discriminate active from inactive molecules.

11.3   METABOLITE BIOMARKER DISCOVERY

Reference [164] considered selecting metabolite biomarkers using jumping emerging patterns for human hepatocellular carcinoma (HCC, a type of cancer) using metabolitic data. Metabolites are used as features, and Matthews Correlation Coefficient[1] was used to rank and select the most promising pairs of metabolites and value intervals. The discovered jumping emerging patterns (JEPs) are useful not only for indicating which combinations of metabolites occur often for disease samples but also for helping with the construction of improved diagnostic models.

---

[1]https://en.wikipedia.org/wiki/Matthews_correlation_coefficient

The paper used the JEPs to identify complex metabolic signatures of HCC. It also studied the relationship between the JEPs and the metabolic pathways.

11.4 STRUCTURAL ALERTS FOR MOLECULAR TOXICITY

In the pharmaceutical industry, early safety evaluation of candidate molecules is needed before significant investments of time and resources are made [141]. Predictive toxicology, which evaluates such safety by utilizing models based on relationships between chemical structures and toxicological activities, is very appealing. Among different approaches, in silico, data-driven techniques are preferred. One important task for developing such predictive toxicology models involves the discovery and definition of structural alerts.

Reference [141] discussed discovering structural alerts for mutagenicity (toxicity) using a subclass of emerging patterns. The study used a data set of molecules described using molecular graphs. Each vertex of a molecular graph represents an atom, and each edge represents the chemical bond between the two connected atoms. The emerging patterns studied in this paper are combinations of connected molecular fragments that occur more frequently in a toxicity class than in the other class. Another study [181] also considered automating knowledge discovery for toxicity prediction using jumping emerging pattern mining.

11.5 IDENTIFYING DISEASE SUBTYPES, AND DISEASE TREATMENT PLANNING

For classifying medical data for patient diagnosis and so on, simple rules and patterns are more desirable than models involving non-linear distance or kernel functions. This is because rules and patterns can help us understand more about the application in addition to performing an accurate classification.

Reference [117] considered discovering novel emerging patterns (rules) that describe (characterize) the gene-expression profiles of more than six subtypes of acute lymphoblastic leukemia patients. Having these patterns to characterize the subtypes is important, since childhood leukemia is a heterogeneous disease containing multiple subtypes, and since different subtypes respond differently to the same therapy. Reference [225] discussed analysis of gene-expression profiles based on emerging patterns, including (a) diagnosis of disease states, (b) derivation of disease treatment plans, and (c) understanding of gene-interaction networks. Reference [59] gave a unified discussion of the results of the above two papers.

11.6 SAFETY AND STREET CRIME ANALYSIS

Since street crimes have a large impact on both the quality of life and human safety, reducing such crimes is an important issue. It has been empirically known that street crime tends to occur not randomly but in particular places of urban areas. That is, incidents of street crime seem to be highly related to the spatial attributes of locations on the streets. Reference [191] performed street crime analysis (on data from Kyoto, Japan), to find emerging patterns to answer this question:

In what kinds of environments are some types of crimes likely to happen?

This study considered attributes that affect natural surveillance, in other words, how easy it is for people (who live or work inside buildings and stores, or who walk on the street) to watch the surroundings. Such attributes include openness of the location, sizes of doors and windows on walls adjacent to the location, building types (whether occupied or not at key time points), number of pedestrians at the location, and density of stores and bars at the location. The most influential emerging patterns found ([191, Table 3]) indicate the following: (1) Criminal activity is more likely near non-wall buildings, where there are few people during the day, and near office buildings, where there are few people at night. (2) Criminal activity is less likely when there are fewer pedestrians from train stations and there are more residential houses or apartments with certain wall components. Therefore, in the area studied, a wall of residential buildings with certain components serves as protection against bag-snatching.

11.7   CHARACTERIZING MUSIC FAMILIES

Comparing collections of music is a general computational approach which has potential for analyzing large repertoires of music [47]. Reference [151] noted that comparing music families is useful for exploratory corpus-level music analysis, and contrast patterns for the families are useful as they describe significant differences between collections of music. More specifically, [151] performed contrast mining and presented some discovered emerging "interval" patterns distinguishing several families of Cretan folk tunes. The study used sequences of features, namely melodic interval in semitones and duration ratios, to represent music segments. It reported emerging patterns (including jumping ones) that appear frequently in some families of Cretan folk tunes but very infrequently (sometimes never) in other families. These patterns show what are highly unique, and hence can be considered as signature music patterns, for different families of musical tunes.

Reference [46] studied antipattern discovery in folk tunes. Antipatterns are basically jumping or almost jumping emerging patterns; they are "general patterns that are rare or even entirely absent from a (given) set of pieces" and "that are frequent in a background set."

11.8   IDENTIFYING INTERACTION TERMS: ADVERSE DRUG REACTION ANALYSIS

Reference [172] developed an emerging pattern (EP)-based approach for identifying and incorporating candidate confounding interaction terms into regression analysis. The aim was to improve adverse drug reaction prediction using longitudinal observational data. The paper considered six drug families that are commonly associated with myocardial infarction in observational healthcare data, where the causal relationship ground truth is known (adverse drug reaction or not). It used EPs that involve drugs and medical events that are associated with the development of myocardial infarction as candidate confounding interaction terms. The authors then selected some of these candidate terms for use in regularized Cox regression. One Cox regression model with elastic net regularization correctly ranked the drug families known to be true adverse drug reactions above those that are not. This was not the case without the inclusion of EPs as candidate confounding interaction terms. The methodology was found to be efficient, it can identify high-order confounding interactions, and it does not require expert input to specify outcome specific confounders. The authors suggested that the methodology can be applied for any outcome of interest to quickly improve the prediction model.

11.9 COUPLED HIDDEN MARKOV MODEL FOR CRITICAL PATIENT CARE

Reference [88] studied the use of contrast patterns as features to build hidden Markov model (HMM) models for early prediction of critical-care patient events like sepsis or septic shock. Sequential contrast patterns were extracted from multiple physiological variables (multi-dimensional time series) on three noninvasive waveform measurements: mean arterial pressure levels, heart rates, and respiratory rates. They were then used as features to represent patient data over time. Coupled hidden Markov models (CHMM) were then built on the pattern defined features. Performance of the multi-channel patterns based, coupled HMM (MCP-HMM) was found to be statistically better than baseline models (SVM and HMM on the original multi-dimensional numerical sequences).

11.10 POSE-BASED HUMAN ACTIVITY RECOGNITION

Reference [205] considered human-action recognition in videos by combining emerging patterns (EPs) with spatial-temporal structure modeling of human poses. A total of 14 joints (head, neck, (left/right)-hand/elbow/shoulder/hip/knee/foot) were used to represent a human pose (see Figure 11.1). These joints were grouped into five body parts (two arms, two legs, and head). Each joint in a video frame was described using five spatial and orientational features. EPs were used to characterize the spatial-temporal structures of human actions, obtained by mining the EPs from sets of poses associated with different actions. The EPs were found to capture the spatial configurations of body parts in one video frame (by spatial-part-sets) as well as the body part movements (by temporal-part-sets). The EPs were characteristics of human actions. The EPs were interpretable, compact, and also robust to errors associated with joint estimations. The EPs were then used as features to derive action-recognition models. Experimental results showed that the approach outperforms state-of-the-art action recognizers on the UCF sport, the Keck Gesture, and the MSR-Action3D datasets.

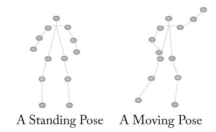

A Standing Pose   A Moving Pose

Figure 11.1: Representing human body using edges linking 14 joints.

11.11 PROTEIN COMPLEX DETECTION

Most protein complex detection methods utilize unsupervised techniques to cluster densely connected nodes in a protein-protein interaction (PPI) network in spite of the fact that many true
complexes are not dense subgraphs. Supervised methods have been proposed recently, but they
do not answer why a group of proteins are predicted as a complex, and they have not inves-
tigated how to detect new complexes of one species by training the model on the PPI data
of another species. Reference [127] proposed an emerging pattern based method to perform
protein complex detection. The method has three major steps: Step 1 uses a feature vector to
represent individual protein complexes. The positive class of data are known complexes and the
negative class are random graphs. Step 2 discovers EPs that distinguish the two classes. Step 3
uses an EP-based (CAEP-like) scoring method to predict if an input is a protein complex. The
method was tested on eight benchmark PPI datasets, and it was compared with seven unsupervised methods, two supervised, and one semi-supervised methods under five standards to assess
the quality of the prediction methods. The results show that in most cases the new method
achieved a better performance, sometimes significantly.

11.12 INHIBITOR PREDICTION COMBINING FCA AND JEP

Reference [17] gave a hybrid classification approach that combines FCA (formal concept analysis) and emerging patterns for the classification of biological inhibitors using their molecule
structures. First, graph mining was applied to the molecular graphs to extract frequent sub-
structures. Then, these sub-structures were used as attributes in a formal context where objects
are molecules of the training set. This formal context was augmented so that each molecule in the
training set has a class defined by its binding mode. A concept lattice was built from the formal
concepts. Moreover, the class information is used for characterizing the concepts whose extents
include objects of a single class or binding mode. The intents of these particular concepts are
JEPs. The last step involved a hierarchical agglomerative clustering process. Based on the knowledge of JEPs and of functional groups, inhibitors are represented as vectors where components
are filled with functional groups and JEPs. The cosine similarity was used for building a den-

drogram which is used for explaining the proximity of some inhibitors and for predicting the binding mode of inhibitors for which this information is still unknown. This hybrid approach was used in an experiment in chemistry for classifying inhibitors of the c-Met protein which plays an important role in protein interactions and in the development of cancer.

11.13 INSTANT ACTIVITY RECOGNITION IN VIDEO SEQUENCES

Reference [207] presented an efficient method to recognize actions in a video sequence from only a couple of frames in real time. The paper employed two types of computationally efficient but perceptually important features, namely optical flow and Canny edge, to capture motion and shape/structure information. The two types of features are known to be non-sparse and can be unreliable or ambiguous at certain parts of videos. To discover strong discriminative information from such features, the paper extended an EP mining method to identify patterns from videos to differentiate action classes. Experimental results showed that the combination of the two types of features achieve superior performance in differentiating actions than using each single type of features alone. The learned patterns are discriminative, statistically significant and reliable, and display semantically meaningful shape-motion structures of human actions. Besides being able to perform instant action recognition, the paper also extended the proposed approach to anomaly detection and sequential-event detection.

11.14 BIRTH DEFECT DETECTION

Reference [219] used emerging patterns (EPs) as features to represent data about child births and addressed the birth defect-detection problem. The original child birth data contained eight features, namely birth weight, birth height, number of births by mother, number of pregnancies by mother, length of pregnancy, mother's age, father's age, and mother's occupation. After mining the EPs, only EPs were used as features to represent the data, and the new data was fed to a decision-tree (C4.5) algorithm. Experimental results showed that the resulting classifier achieved high accuracy, and it outperformed other well-known classification algorithms.

11.15 SURGERY STAGE IDENTIFICATION AND FEEDBACK DELIVERY

Reference [234] proposed a framework based on emerging patterns to evaluate surgical performance and provide feedback during simulated ear (temporal bone) surgery in a 3D virtual environment. Temporal bone surgery is composed of a number of stages with distinct aims and surgical techniques. To provide context-appropriate feedback it is imperative to be able to identify each stage, recognize when feedback is to be provided, and determine the nature of that feedback. To achieve these aims, several CAEP-based models were trained using data recorded by a temporal bone simulator. One model is used to predict the current stage of the procedure, and separate stage-specific models are used to provide human-friendly feedback within each stage. Experiments showed that the proposed system identifies the stage of the procedure correctly and provides constructive feedback to assist surgical trainees in improving their technique.

11.16 SENSOR-BASED ACTIVITY RECOGNITION

Several researchers considered using EPs for complex activity recognition in the smart-home senior-care environment using sensor data. A simple activity is performed by a single person during which no other activity is carried out concurrently. A complex activity includes interleaved and parallel actions that are parts of several simple activities. Reference [90] used emerging patterns to derive a likelihood score (in the CAEP aggregation style) for each window in a sequence of activities. The paper found that this score by itself already has a fairly good performance, and adding other features helped improve the performance. Reference [136] used EPs to derive a different likelihood score (also in the CAEP aggregation style) for each window in a sequence of activities; the contribution of an EP P (having C_i as its home activity, viewed as a class) matching a window is given by $\mathsf{supp}(X, D) * dpower(X, C_i)$, where

$$dpower(X, C_i) = \Pr(C_i \mid X) * \Pr(not(C_i) \mid not(X)),$$

with $not(X)$ representing "X is not satisfied" and $not(C_i)$ representing "not in class C_i." (The *dpower* measure was first defined in [89], and later used in [136].) Then the scores were used as features. Combining these features with Random Forest led to the highest accuracy among all classifiers considered.

11.17 ONLINE BANKING FRAUD DETECTION

Reference [213] studied the detection of sophisticated online banking fraud on extremely imbalanced data. It proposed a system that combines three risk scores predicted by three models, one of which being the CAEP-style likelihood score. Results from experiments on large-scale real online banking data demonstrate that the proposed system can achieve substantially higher accuracy and lower alert volume than the latest benchmark fraud-detection system (which incorporates domain knowledge and traditional fraud-detection methods). Observe that having higher accuracy and lower alert volume is very important to the user, as this implies that more frauds are caught and fewer false alarms are raised. Each false alarm alert must be processed manually involving much wasted human time and effort.

11.18 OTHER EP-BASED CLASSIFICATION APPROACHES AND STUDIES

Many emerging pattern based classifiers have been introduced. Below we highlight several that have highly distinctive techniques, and list the others. Some other approaches were covered in Chapters 4 and 5 and are not repeated here.

Reference [70] introduced an approach to combine EPs with the Naive Bayes classification method. It limits EPs to "essential EPs," defined as EPs with high growthRate and fairly high supp, and they are minimal in the collection of EPs (for some given support and growthRate thresholds). It ranks the EPs P using $\mathsf{supp}(P, C_i) * \frac{\mathsf{growthRate}(P,C_i)}{\mathsf{growthRate}(P,C_i)+1}$. For each instance t, the set of matching EPs is selected from the ranked order one at a time, and a new EP is selected only if it contains some item in t not already contained in the set of selected EPs. The set of selected EPs are used to compute the probabilities of t in each given class in the usual Naive Bayes manner. The method was found to produce good classification results. This method uses fewer EPs to classify test instances than CAEP. The classifier uses aggregated votes (as probabilistic products) of matching patterns to classify instances.

Reference [72] used CAEP's likelihood scores to constructed weighted support vector machines (SVMs). The proposed method used CAEP's likelihood scores as weights for the training data instances. These weights reflect "confidence" of membership of the instances in their assigned classes. The weights were then used by SVM in the learning process. The proposed method was found to outperform traditional SVM. Reference [12] considered building weighted decision trees. Reference [230] considered EP-based robust weighting scheme for fuzzy SVMs.

Reference [82] introduced a fuzzy emerging pattern based method for classification. Fuzzy emerging patterns are an extension of emerging patterns to deal with numerical attributes using fuzzy discretization. The classifier introduced in this paper was shown to outperform some popular and state-of-the-art classifiers on several UCI repository datasets. The classifier uses aggregated votes to classify instances.

Reference [232] used statistics of CAEP's likelihood scores to improve CAEP's classification. Reference [118] considered combining the strength of pattern frequency and distance for classification. Reference [21] studied classification using constrained emerging patterns. Reference [211] studied exploiting "maximal" emerging patterns for classification. Reference [71] studied noise tolerant classification by "chi" emerging patterns. Reference [10] studied using emerging patterns to create new instances for imbalanced classification. Reference [9] considered using emerging patterns and decision trees in rare-class classification. Reference [8] considered using emerging patterns for improving the quality of rare-class classification. Reference [68] considered further improving emerging pattern-based classifiers via bagging. Reference [129] introduced an emerging pattern-based classification approach for graph classification. Reference [16] gave a lazy approach to privacy-preserving classification with emerging patterns. Reference [83] gave a survey of emerging pattern-based approaches for supervised classification. Reference [161] presented a classification algorithm based on emerging patterns. Reference [134] presented a new contrast pattern-based classifier for class imbalance problems. Reference [13] evaluated performance of some emerging pattern-based classifiers, using emerging patterns computed by an ant-colony-based algorithm. Reference [43] proposed a new contrast pattern-based classification method for imbalanced data, which incorporates considerations on pattern quality and class imbalance ratio. Reference [197] proposed an adaptive classification method using jumping emerging patterns. Reference [132] proposed to select all the contrast patterns for the minority class and a certain percent of contrast patterns for the majority class in order to address imbalanced classification; the method was found to outperform other classification methods for detecting pneumatic failures on temporary immersion bioreactors[2] in [133].

11.19 EMERGING PATTERNS FOR CLASSIFICATION OVER STREAMING DATA

Many datasets from real-world applications have very high-dimensional or increasing feature space. It is a challenge to learn and maintain a classifier to deal with very high dimensionality or streaming features. Reference [227] handled this challenge by proposing a semi-streaming approach (called EPSF) to maintain the set of streaming features and to perform emerging pattern mining for a recent temporal window, and also adapting emerging-pattern-based classification models. EPSF builds two pools when processing streaming data: a feature pool and a 1-itemset EP pool, and periodically computes and updates emerging patterns formed from the 1-itemset EP pool for the construction and maintenance of an EP classification model. For efficiency and accuracy purposes, it uses iCAEP's approach [231] to classify instances.

Reference [11] also considered mining emerging patterns for classification over streaming data. Reference [69] proposed an efficient single-scan algorithm for mining "essential" jumping emerging patterns for classification. Reference [81] considered pattern types related to emerging patterns, namely abrupt emergences of episodes and abrupt submergences of episodes, for monitoring data streams.

11.20 OTHER STUDIES AND APPLICATIONS

The references cited in this section are very diverse. Although it is hard to do, this section is still organized into some kind of groupings whenever possible.

This group of papers is related to sequence data, text data, and graph data. Reference [187] used contrast mining on genotype data to discover novel gene associations specific to autism subgroups. Reference [233] used contrast sequence patterns in taxpayer behavior analysis. Reference [50] considered using emerging subsequences as features; it proposed to use a sliding-window-based mechanism for matching patterns with sequences; experiments confirmed that emerging subsequence features outperform frequent-subsequence features by up to 20% in accuracy. Reference [87] considered risk prediction for acute hypotensive patients by using gap

---

[2]Temporary immersion bioreactors are used to increase plant quality and plant multiplication rates.

constrained sequential contrast patterns. Reference [178] considered peptide folding prediction by using contrast patterns as features. Experimental results on two benchmark protein datasets indicated that these features can outperform other approaches. References [99, 199] presented an emerging pattern-based method called PolyA-iEP for the effective prediction of polyadenylation sites. Reference [224] used emerging patterns in the development of a minimally supervised method for multilingual paraphrase extraction from definition sentences on the Web. Reference [165] considered using emerging patterns to describe and analyze graph topological patterns that indicate co-variations among vertex descriptors.

This group is related to image classification. Reference [203] used histograms of jumping emerging patterns for image classification and object recognition. Reference [3] proposed using emerging patterns as features for image classification. This approach was found to reduce dimensionality by over 50% compared with the frequent pattern approach, and to produce better classification results.

This group is broadly related to discovery of trends or niche opportunities, or exploration of hypothesis. Reference [111] used emerging pattern mining to identify emerging hotel preferences. Reference [53] used emerging patterns to perform niche opportunity discovery. Reference [123] utilized and extended contrast mining to support exploratory hypothesis testing and analysis.

This group is related to disaster and risk analysis. Reference [166] used contrast patterns as features for intent classification of short-text data on social media in the context of disaster analysis. Reference [171] used fuzzy emerging patterns for modeling landslides' susceptibility. References [109, 159] used emerging pattern mining for identifying safe and non-safe electric power lines and electric power management.

This group is related to sport analysis. Reference [36] used growthRate of patterns to evaluate deviant gaming behavior and to analyze where a player erred in MOBA games. Reference [64] used emerging patterns to analyze why sports officials drop out.

This group is related to traffic analysis. Reference [40] used contrast mining to summarize significant changes in network traffic. Reference [210] considered analyzing the impact of urban traffic interventions using contrast mining on vehicle trajectory data. Reference [209] used emerging pattern for traffic trajectory analysis in the context of the Internet of Things.

This group is related to retail and business analytic problems. Reference [124] studied the mining of textual customer reviews in order to discover why customers like or dislike certain products and services. Reference [184] studied the use of emerging patterns to analyze the change of customer behavior in an Internet shopping mall. Reference [215] considered the analysis of sales trends by mining emerging patterns in dynamic markets. Reference [155] reported a study that used EPs to characterize events that lead to large negative return of stocks after downgrade events. Reference [7] considered contrast analysis techniques for recommendation; it defined sets of contrasting rules as a pattern and used them to identify trigger factors "that can stimulate the transition of data elements from one class to another." Reference [156] discussed capabilities of contrast mining application in SWOT analysis. SWOT analysis is a strategic planning technique used to help users to identify strengths, weaknesses, opportunities, and threats related to business competition or project planning. Reference [146] used emerging patterns to analyze high- vs. low-fashion sensitivity based on factors such as personal color preferences (color psychology) over different seasons.

This group is related to medicine and healthcare informatics. Using contrast mining, [169] studied the discovery of multifactorial patterns that are strongly related to the development of age-related cataract using a large volume of electronic health records (EHR). The researchers wanted to understand the cumulative effect of multiple factors along with clinical and systemic disease conditions. References [107, 162, 176] used emerging patterns and also CAEP for diagnosing diseases in the medical domain.

This group is related to bioinformatics and chemoinformatics. Reference [163] considered discovering emerging graph patterns from chemicals. Reference [101] considered emerging pattern-based subspace clustering for microarray gene-expression data. Reference [139] used emerging patterns to evaluate inter-laboratory and cross-platform concordance of microarray gene-expression technologies.

This group is related to detecting unusual activities. Reference [96] used emerging patterns to detect unusual activities for temporal debugging of embedded streaming applications. Reference [177] used contrast mining for the detection of illicit behavior.

Reference [45] considered privacy-preserving data publishing in the context of frequent itemsets mining; it studied how to hide emerging patterns in the original data by a minimal data distortion.

Reference [102] examined the broad topic of "difference analysis in Big Data." Reference [29] considered emerging patterns in interpretation analysis.

11.21 SUMMARY OF USES: APPLICATION DOMAIN PERSPECTIVE

This section only discusses directions that have a substantial number of published papers.

Emerging patterns have been used often in chemoinformatics research. In [18], emerging patterns and the CAEP methodology were adapted in chemoinformatics as emerging chemical patterns (ECP) for classification of active compounds. Later, ECP was applied to simulate iterative screening experiments [20] and to analyze bioactive compound conformations [19]. More recently, ECP was applied to classify compounds with multi-target activities [149], to identify compounds forming activity cliffs [150], and to identify compounds having different local structure-activity relationships [148]. Reference [135] used jumping fragments in combination with QSARs for the assessment of classification in ecotoxicology. Reference [140] considered exploring structure–activity relationships of chemicals. References [181, 182] used jumping emerging patterns for automating knowledge discovery and for toxicity analysis and prediction. Reference [110] provided a survey on automated detection of structural alerts (chemical fragments) in (eco) toxicology. Reference [141] considered discovering structural alerts for mutagenicity using stable emerging molecular patterns. Reference [48] used emerging pattern mining techniques to discover new structural alerts for Ames mutagenicity.

There have also been many studies in the fields of bioinformatics, medicine, and healthcare. These include gene ranking for complex diseases (see Chapter 5), disease subtype analysis [117], disease diagnosis [107, 162, 176, 225], biomarker discovery [164], adverse drug identification [172], critical patient care [88], protein complex detection [127], inhibitor prediction [17], birth defect detection [219], surgery stage identification and feedback delivery [234], risk prediction for acute hypotensive patients [87], peptide-folding prediction [178], polyadenylation site prediction [99, 199], prognostic risk models for heart failure-patient survival [194], and treatment planning for traumatic-brain-injury patients [193].

There have also been quite a few studies concerning activity recognition, including human-action recognition in videos [205, 207] and sensor-based activity recognition [89, 90, 136].

11.22 DISCUSSION

Perhaps the most representative use of emerging patterns is centered around data group characterization. Such characterization is fundamental for understanding data groups, discovering important attributes, constructing pattern-based features, and so on.

It should be noted that using emerging patterns as features is often more effective than using frequent patterns as features, as the number of EPs is often smaller, and EPs are closely linked to the data groups/classes of interest/importance.

A recent focus of classification research is to build interpretable predictive models. Reference [106] proposed to use interpretable decision sets involving sets of independent if-then rules as a framework for building predictive models that are highly accurate, yet also highly interpretable. While emerging pattern-based classifiers are somehow interpretable, it is anticipated that further research can make them more interpretable and accurate.

Bibliography

[1] Neda Abdelhamid, Aladdin Ayesh, and Fadi Thabtah. Phishing detection based associative classification data mining. *Expert Systems with Applications*, 41(13):5948–5959,
2014. DOI: 10.1016/j.eswa.2014.03.019 36

[2] Pierre Accorsi, Arnaud Sallaberry, Nathalie Lalande, Sandra Bringay, Florence Le Ber,
Pascal Poncelet, Mickaël Fabrègue, Flavie Cernesson, Agnès Braud, and Maguelonne
Teisseire. Hydroqual: Visual analysis of river water quality. In *VAST: Visual Analytics
Science and Technology*, pages 123–132, 2014. DOI: 10.1109/vast.2014.7042488 18

[3] Niusvel Acosta-Mendoza, Andrés Gago-Alonso, Jesús Ariel Carrasco-Ochoa, José Francisco Martínez-Trinidad, and José Eladio Medina-Pagola. Improving graph-based image
classification by using emerging patterns as attributes. *Engineering Applications of Artificial Intelligence*, 50:215–225, 2016. DOI: 10.1016/j.engappai.2016.01.030 97

[4] Charu C. Aggarwal and Philip S. Yu. Outlier detection for high dimensional data. In
ACM SIGMOD Record, vol. 30, pages 37–46, 2001. DOI: 10.1145/376284.375668 44

[5] Rakesh Agrawal, Tomasz Imieliński, and Arun Swami. Mining association rules between
sets of items in large databases. In *ACM SIGMOD Conference*, pages 207–216, 1993.
DOI: 10.1145/170036.170072 12, 22

[6] Javier Lopez Alberca. Natural emerging patterns: A reformulation for classification.
Ph.D. thesis, Royal Institute of Technology, 2014. 25

[7] Marharyta Aleksandrova. Matrix factorization and contrast analysis techniques for recommendation. Ph.D. thesis, Université de Lorraine, 2017. 97

[8] Hamad Alhammady and Kotagiri Ramamohanarao. The application of emerging patterns for improving the quality of rare-class classification. In *Pacific-Asia Conference on Knowledge Discovery and Data Mining*, pages 207–211, Springer, 2004. DOI:
10.1007/978-3-540-24775-3_27 95

[9] Hamad Alhammady and Kotagiri Ramamohanarao. Using emerging patterns and decision trees in rare-class classification. In *Proc. of IEEE International Conference on Data
Mining (ICDM)*, pages 315–318, 2004. DOI: 10.1109/icdm.2004.10058 95

[10] Hamad Alhammady and Kotagiri Ramamohanarao. Expanding the training data space using emerging patterns and genetic methods. In *Proc. of SIAM International Conference on Data Mining*, pages 481–485, 2005. DOI: 10.1137/1.9781611972757.45 95

[11] Hamad Alhammady and Kotagiri Ramamohanarao. Mining emerging patterns and classification in data streams. In *Proc. of IEEE/WIC/ACM International Conference on Web Intelligence*, pages 272–275, 2005. DOI: 10.1109/wi.2005.96 96

[12] Hamad Alhammady and Kotagiri Ramamohanarao. Using emerging patterns to construct weighted decision trees. *IEEE Transactions on Knowledge and Data Engineering*, 18(7):865–876, 2006. DOI: 10.1109/tkde.2006.116 95

[13] Zulfiqar Ali and Waseem Shahzad. EPACO: A novel ant colony optimization for emerging patterns based classification. *Cluster Computing*, pages 1–15, 2017. DOI: 10.1007/s10586-017-0894-4 96

[14] U. Alon, N. Barkai, et al. Broad patterns of gene expression revealed by clustering analysis of tumor and normal colon tissues probed by oligonucleotide arrays. *Proc. of the National Academy of Science*, 96:6745–6750, 1999. DOI: 10.1073/pnas.96.12.6745 72

[15] William F. Anderson and Rayna Matsuno. Breast cancer heterogeneity: A mixture of at least two main types? *Journal of the National Cancer Institute*, 98:948–951, 2006. DOI: 10.1093/jnci/djj295 83

[16] Piotr Andruszkiewicz. Lazy approach to privacy preserving classification with emerging patterns. In *Proc. of the 19th International Symposium of Emerging Intelligent Technologies in Industry (ISMIS)*, pages 253–268, 2011. DOI: 10.1007/978-3-642-22732-5_21 95

[17] Yasmine Asses, Aleksey Buzmakov, Thomas Bourquard, Sergei O. Kuznetsov, and Amedeo Napoli. A hybrid classification approach based on FCA and emerging patterns-an application for the classification of biological inhibitors. In *CLA: The 9th International Conference on Concept Lattices and their Applications*, 2012. 8, 92, 99

[18] Jens Auer and Jurgen Bajorath. Emerging chemical patterns: A new methodology for molecular classification and compound selection. *Journal of Chemical Information and Modeling*, 46(6):2502–2514, 2006. DOI: 10.1002/chin.200708200 30, 33, 37, 38, 39, 87, 98

[19] Jens Auer and Jurgen Bajorath. Distinguishing between bioactive and modeled compound conformations through mining of emerging chemical patterns. *Journal of Chemical Information and Modeling*, 48(9):1747–1753, 2008. DOI: 10.1021/ci8001793 88, 98

[20] Jens Auer and Jurgen Bajorath. Simulation of sequential screening experiments using emerging chemical patterns. *Medicinal Chemistry*, 4(1):80–90, 2008. DOI: 10.2174/157340608783331452 88, 98

[21] James Bailey, Thomas Manoukian, and Kotagiri Ramamohanarao. Classification using constrained emerging patterns. In *International Conference on Web-Age Information Management*, pages 226–237, Springer, 2003. DOI: 10.1007/978-3-540-45160-0_22 95

[22] James Bailey, Thomas Manoukian, and Kotagiri Ramamohanarao. A fast algorithm for computing hypergraph transversals and its application in mining emerging patterns. In *Proc. of IEEE International Conference on Data Mining (ICDM)*, pages 485–488, 2003. DOI: 10.1109/icdm.2003.1250958 25

[23] Elena Baralis, Silvia Chiusano, and Paolo Garza. A lazy approach to associative classification. *IEEE Transactions on Knowledge and Data Engineering*, 20(2):156–171, 2008. DOI: 10.1109/tkde.2007.190677 36

[24] Yves Bastide, Rafik Taouil, Nicolas Pasquier, Gerd Stumme, and Lotfi Lakhal. Mining frequent patterns with counting inference. *ACM SIGKDD Explorations Newsletter*, 2(2):66–75, 2000. DOI: 10.1145/380995.381017 13

[25] Stephen D. Bay and Michael J. Pazzani. Detecting change in categorical data: Mining contrast sets. In *ACM International Conference on Knowledge Discovery and Data Mining (KDD)*, pages 302–306, 1999. 2, 8, 17 DOI: 10.1145/312129.312263

[26] Stephen D. Bay and Michael J. Pazzani. Characterizing model errors and differences. In *Proc. of International Conference on Machine Learning (ICML)*, pages 49–56, 2000. 8

[27] Stephen D. Bay and Michael J. Pazzani. Detecting group differences: Mining contrast sets. *Data Mining Knowledge Discovery*, 5(3):213–246, 2001. 8, 17

[28] Kevin Beyer, Jonathan Goldstein, Raghu Ramakrishnan, and Uri Shaft. When is "nearest neighbor" meaningful? In *International Conference on Database Theory*, pages 217–235, Springer, 1999. DOI: 10.1007/3-540-49257-7_15 44, 58

[29] Philipp Blandfort, Jörn Hees, and Desmond U. Patton. An overview of computational approaches for analyzing interpretation. *ArXiv Preprint ArXiv:1811.04028*, 2018. 98

[30] Ilona Bluemke and Marcin Tarka. Detection of anomalies in a SOA system by learning algorithms. In *Complex Systems and Dependability*, pages 69–85, Springer, 2013. DOI: 10.1007/978-3-642-30662-4_5 50

[31] Anne-Laure Boulesteix and Gerhard Tutz. Identification of interaction patterns and classification with applications to microarray data. *Computational Statistics and Data Analysis*, 50(3):783–802, 2006. DOI: 10.1016/j.csda.2004.10.004 69

[32] Evan A. Boyle, Yang I. Li, and Jonathan K. Pritchard. An expanded view of complex traits: From polygenic to omnigenic. *Cell*, 169(7):1177–1186, 2017. DOI: 10.1016/j.cell.2017.05.038 68

[33] Björn Bringmann, Siegfried Nijssen, and Albrecht Zimmermann. Pattern-based classification: A unifying perspective. *ArXiv Preprint ArXiv:1111.6191*, 2011. 36

[34] Anna L. Buczak and Erhan Guven. A survey of data mining and machine learning methods for cyber security intrusion detection. *IEEE Communications Surveys and Tutorials*, 18(2):1153–1176, 2016. DOI: 10.1109/comst.2015.2494502 44

[35] Cristóbal J. Carmona, M. J. del Jesus, and Francisco Herrera. A unifying analysis for the supervised descriptive rule discovery via the weighted relative accuracy. *Knowledge-Based Systems*, 139:89–100, 2018. DOI: 10.1016/j.knosys.2017.10.015 8

[36] Olivier Cavadenti, Victor Codocedo, Jean-François Boulicaut, and Mehdi Kaytoue. What did I do wrong in my MOBA game? Mining patterns discriminating deviant behaviours. In *IEEE International Conference on Data Science and Advanced Analytics (DSAA)*, pages 662–671, 2016. DOI: 10.1109/dsaa.2016.75 18, 97

[37] Michelangelo Ceci, Annalisa Appice, Costantina Caruso, and Donato Malerba. Discovering emerging patterns for anomaly detection in network connection data. In *International Symposium on Methodologies for Intelligent Systems*, pages 179–188, Springer, 2008. DOI: 10.1007/978-3-540-68123-6_20 50

[38] Michelangelo Ceci, Annalisa Appice, Lucrezia Macchia, and Donato Malerba. Relational classification based on emerging patterns. In *Proc. of the 16th Italian Symposium on Advanced Database Systems (SEBD)*, pages 45–56, 2008. 33

[39] Chih-Chung Chang and Chih-Jen Lin. LIBSVM: A library for support vector machines. *ACM Transactions on Intelligent Systems and Technology (TIST)*, 2(3):27, 2011. DOI: 10.1145/1961189.1961199 48

[40] Elaheh Alipour Chavary, Sarah M. Erfani, and Christopher Leckie. Summarizing significant changes in network traffic using contrast pattern mining. In *Proc. of ACM Conference on Information and Knowledge Management*, pages 2015–2018, 2017. DOI: 10.1145/3132847.3133111 97

[41] Lijun Chen and Guozhu Dong. Masquerader detection using OCLEP: One class classification using length statistics of emerging patterns. In *International Workshop on Information Processing over Evolving Networks (WINPEN)*, 2006. DOI: 10.1109/waimw.2006.19 45, 49

[42] Lijun Chen and Guozhu Dong. Using emerging patterns in outlier and rare-class prediction. In *Contrast Data Mining: Concepts, Algorithms, and Applications*, Guozhu Dong and James Bailey, Eds., pages 171–186, Chapman & Hall/CRC, 2012. DOI: 10.1201/b12986-17 49

[43] Xiangtao Chen, Yajing Gao, and Siqi Ren. A new contrast pattern-based classification for imbalanced data. In *Proc. of the 2nd International Symposium on Computer Science and Intelligent Control*, page 45, ACM, 2018. DOI: 10.1145/3284557.3284708 96

[44] Hong Cheng, Xifeng Yan, Jiawei Han, and Philip S. Yu. Direct discriminative pattern mining for effective classification. In *Proc. of IEEE International Conference on Data Engineering*, pages 169–178, 2008. DOI: 10.1109/icde.2008.4497425 8, 25

[45] Michael W. K. Cheng, Byron Choi, and William Kwok-Wai Cheung. Hiding emerging patterns with local recoding generalization. In *Proc. of Advances in Knowledge Discovery and Data Mining (PAKDD)*, pages 158–170, 2010. DOI: 10.1007/978-3-642-13657-3_19 98

[46] Darrell Conklin. Antipattern discovery in folk tunes. *Journal of New Music Research*, 42(2):161–169, 2013. DOI: 10.1080/09298215.2013.809125 90

[47] Nicholas Cook. Computational and comparative musicology. *Empirical Musicology: Aims, Methods, Prospects*, pages 103–126, 2004. DOI: 10.1093/acprof:oso/9780195167498.003.0006 90

[48] Laurence Coquin, Steven J. Canipa, William C. Drewe, Lilia Fisk, Valerie J. Gillet, Mukesh Patel, Jeffrey Plante, Richard J. Sherhod, and Jonathan D. Vessey. New structural alerts for ames mutagenicity discovered using emerging pattern mining techniques. *Toxicology Research*, 4(1):46–56, 2015. DOI: 10.1039/c4tx00071d 99

[49] Val Curtis, Robert Aunger, and Tamer Rabie. Evidence that disgust evolved to protect from risk of disease. *Proc. of the Royal Society of London B: Biological Sciences*, 271(Suppl 4):S131–S133, 2004. DOI: 10.1098/rsbl.2003.0144 2

[50] Kang Deng and Osmar R. Zaïane. Contrasting sequence groups by emerging sequences. In *International Conference on Discovery Science*, pages 377–384, Springer, 2009. DOI: 10.1007/978-3-642-04747-3_29 96

[51] Guozhu Dong and James Bailey. *Contrast Data Mining: Concepts, Algorithms, and Applications*. CRC Press, 2012. DOI: 10.1201/b12986 25

[52] Guozhu Dong and Sanjeev Bhatta. Subpopulation-wise conditional correlation modeling and analysis. In *IEEE International Conference on Big Knowledge (ICBK)*, pages 17–24, 2017. DOI: 10.1109/icbk.2017.19 85

[53] Guozhu Dong and Kaustubh Deshpande. Efficient mining of niches and set routines. In *Proc. of Pacific-Asia Conference on Knowledge Discovery and Data Mining (PAKDD)*, pages 234–246, 2001. DOI: 10.1007/3-540-45357-1_27 97

[54] Guozhu Dong and Neil Fore. Discovering dynamic logical blog communities based on their distinct interest profiles. In *Proc. of International Conference on Social Eco-Informatics (SOTICS)*, 2011. 63

[55] Guozhu Dong, Chunyu Jiang, Jian Pei, Jinyan Li, and Limsoon Wong. Mining succinct systems of minimal generators of formal concepts. In *Database Systems for Advanced Applications (DASFAA)*, pages 175–187, 2005. DOI: 10.1007/11408079_17 13

[56] Guozhu Dong and Jinyan Li. Efficient mining of emerging patterns: Discovering trends and differences. In *Proc. of ACM Conference on Knowledge Discovery and Data Mining (KDD)*, pages 43–52, 1999. DOI: 10.1145/312129.312191 2, 13, 16, 17, 18, 46, 71

[57] Guozhu Dong and Jinyan Li. Mining border descriptions of emerging patterns from dataset pairs. *Knowledge and Information Systems*, 8(2):178–202, 2005. DOI: 10.1007/s10115-004-0178-1 16, 18, 25

[58] Guozhu Dong, Jinyan Li, Guimei Liu, and Limsoon Wong. *Mining Conditional Contrast Patterns. Chapter in Post-Mining of Association Rules: Techniques for Effective Knowledge Extraction.* Yanchang Zhao, Chengqi Zhang, and Longbing Cao, Eds., IGI Global, 2009. 25

[59] Guozhu Dong, Jinyan Li, and Limsoon Wong. *The Use of Emerging Patterns in the Analysis of Gene Expression Profiles for the Diagnosis and Understanding of Diseases.* Chapter in *New Generation of Data Mining Applications*, Mehmed Kantardzic and Jozef Zurada, Eds., IEEE Press, 2005. 89

[60] Guozhu Dong and Sai Kiran Pentukar. OCLEP+: One-class anomaly and intrusion detection using minimal length of emerging patterns. *CSE Tech Report*, Wright State University, available at arXiv:1811.09842, 2018. 49

[61] Guozhu Dong and Vahid Taslimitehrani. Pattern aided regression modeling and prediction model analysis. *IEEE Transactions on Knowledge Data Engineering (TKDE)*, 27:2452–2465, 2015. DOI: 10.1109/icde.2016.7498398 78, 80, 82, 83

[62] Guozhu Dong and Vahid Taslimitehrani. Pattern aided classification. In *Proc. of SIAM International Conference on Data Mining*, pages 225–233, 2016. DOI: 10.1137/1.9781611974348.26 80, 82, 84

[63] Guozhu Dong, Xiuzhen Zhang, Limsoon Wong, and Jinyan Li. CAEP: Classification by aggregating emerging patterns. In *Proc. of Discovery Science*, pages 30–42, 1999. DOI: 10.1007/3-540-46846-3_4 33, 36

[64] Fabrice Dosseville, François Rioult, and Sylvain Laborde.   Why do sports officials dropout? In *Machine Learning and Data Mining for Sports Analytics*, 2013. 97

[65] James Dougherty, Ron Kohavi, and Mehran Sahami. Supervised and unsupervised discretization of continuous features. In *Proc. of International Conference on Machine Learning (ICML)*, pages 194–202, 1995. DOI: 10.1016/b978-1-55860-377-6.50032-3 10, 66

[66] Wouter Duivesteijn and Julia Thaele.   Understanding where your classifier does (not) work.   In *Joint European Conference on Machine Learning and Knowledge Discovery in Databases*, pages 250–253, Springer, 2015. DOI: 10.1109/icdm.2014.10 8

[67] William DuMouchel, Wen-Hua Ju, Alan F. Karr, Matthias Schonlau, Martin Theusan, and Yehuda Vardi. Computer intrusion: Detecting masquerades. *Statistical Science*, 16(1):1–17, 2001. DOI: 10.1214/ss/998929476 48

[68] Hongjian Fan, Ming Fan, Kotagiri Ramamohanarao, and Mengxu Liu. Further improving emerging pattern based classifiers via bagging. In *Pacific-Asia Conference on Knowledge Discovery and Data Mining*, pages 91–96, Springer, 2006. DOI: 10.1007/11731139_13 36, 95

[69] Hongjian Fan and Kotagiri Ramamohanarao. An efficient single-scan algorithm for mining essential jumping emerging patterns for classification. In *Proc. of Pacific-Asia Conference on Knowledge Discovery and Data Mining (PAKDD)*, pages 456–462, 2002. DOI: 10.1007/3-540-47887-6_45 96

[70] Hongjian Fan and Kotagiri Ramamohanarao.   A Bayesian approach to use emerging patterns for classification. In *Proc. of Australasian Database Conference*, pages 39–48, 2003. 25, 36, 95

[71] Hongjian Fan and Kotagiri Ramamohanarao.   Noise tolerant classification by chi emerging patterns. In *Pacific-Asia Conference on Knowledge Discovery and Data Mining*, pages 201–206, Springer, 2004. DOI: 10.1007/978-3-540-24775-3_26 36, 95

[72] Hongjian Fan and Kotagiri Ramamohanarao. A weighting scheme based on emerging patterns for weighted support vector machines. In *Proc. of IEEE International Conference on Granular Computing*, pages 435–440, 2005. DOI: 10.1109/grc.2005.1547329 95

[73] Gang Fang, Majda Haznadar, Wen Wang, Haoyu Yu, Michael Steinbach, Timothy R. Church, William S. Oetting, Brian Van Ness, and Vipin Kumar.   Highorder SNP combinations associated with complex diseases: Efficient discovery, statistical power and functional interactions. *PloS One*, 7(4):e33531, 2012. DOI: 10.1371/journal.pone.0033531 69

[74] Gang Fang, Gaurav Pandey, Wen Wang, Manish Gupta, Michael Steinbach, and Vipin Kumar. Mining low-support discriminative patterns from dense and high-dimensional data. *Knowledge and Data Engineering, IEEE Transactions on*, 24(2):279–294, 2012. DOI: 10.1109/tkde.2010.241 8

[75] Gang Fang, Wen Wang, Benjamin Oatley, Brian Van Ness, Michael Steinbach, and Vipin Kumar. Characterizing discriminative patterns. *Computing Research Repository*, abs/1102.4, 2011. 69

[76] Fabio Fassetti, Simona E. Rombo, and Cristina Serrao. Exceptional pattern discovery. In *Discriminative Pattern Discovery on Biological Networks*, pages 23–30, Springer, 2017. DOI: 10.1007/978-3-319-63477-7_3 8

[77] Usama M. Fayyad and Keki B. Irani. Multi-interval discretization of continuous-valued attributes for classification learning. In *Proc. of the International Joint Conference on Artificial Intelligence (IJCAI)*, pages 1022–1029, 1993. DOI: 10.18297/etd/18 11

[78] Mengling Feng, Guozhu Dong, Jinyan Li, Yap-Peng Tan, and Limsoon Wong. Pattern space maintenance for data updates and interactive mining. *Computational Intelligence*, 26(3):282–317, 2010. DOI: 10.1111/j.1467-8640.2010.00360.x 62

[79] Juan L. Fernández-Martínez, Stephen T. Sonis, et al. Sensitivity analysis of gene ranking methods in phenotype prediction. *Journal of Biomedical Informatics*, 64:255–264, 2016. DOI: 10.1016/j.jbi.2016.10.012 66

[80] Neil Fore and Guozhu Dong. *CPC: A Contrast Pattern Based Clustering Algorithm*. Chapter in *Contrast Data Mining: Concepts, Algorithms and Applications*, Guozhu Dong and James Bailey, Eds., Chapman & Hall/CRC, 2013. 61, 62

[81] Min Gan and Honghua Dai. Detecting and monitoring abrupt emergences and submergences of episodes over data streams. *Information Systems*, 39:277–289, 2014. DOI: 10.1016/j.is.2012.05.009 96

[82] Milton García-Borroto, José Fco Martínez-Trinidad, and Jesús Ariel Carrasco-Ochoa. Fuzzy emerging patterns for classifying hard domains. *Knowledge and Information Systems*, 28(2):473–489, 2011. DOI: 10.1007/s10115-010-0324-x 36, 95

[83] Milton García-Borroto, José Fco Martínez-Trinidad, and Jesús Ariel Carrasco-Ochoa. A survey of emerging patterns for supervised classification. *Artificial Intelligence Review*, 42(4):705–721, 2014. DOI: 10.1007/s10462-012-9355-x 36, 95

[84] Milton García-Borroto, José Fco Martínez-Trinidad, Jesús Ariel Carrasco-Ochoa, Miguel Angel Medina-Pérez, and José Ruiz-Shulcloper. LCMine: An efficient algorithm for mining discriminative regularities and its application in supervised classifica-

tion. *Pattern Recognition*, 43(9):3025–3034, 2010. DOI: 10.1016/j.patcog.2010.04.008 25, 36

[85] A. M. García-Vico, C. J. Carmona, D. Martín, M. García-Borroto, and M. J. del Jesus. An overview of emerging pattern mining in supervised descriptive rule discovery: Taxonomy, empirical study, trends, and prospects. *Wiley Interdisciplinary Reviews: Data Mining and Knowledge Discovery*, 2017. DOI: 10.1002/widm.1231 8

[86] Behzad Ghanbarian, Vahid Taslimitehrani, Guozhu Dong, and Yakov A. Pachepsky. Sample dimensions effect on prediction of soil water retention curve and saturated hydraulic conductivity. *Journal of Hydrology*, 528:127–137, 2015. DOI: 10.1016/j.jhydrol.2015.06.024 84

[87] Shameek Ghosh, Mengling Feng, Hung Nguyen, and Jinyan Li. Risk prediction for acute hypotensive patients by using gap constrained sequential contrast patterns. In *AMIA Annual Symposium Proceedings*, page 1748, American Medical Informatics Association, 2014. 96, 99

[88] Shameek Ghosh, Jinyan Li, Longbing Cao, and Kotagiri Ramamohanarao. Septic shock prediction for ICU patients via coupled HMM walking on sequential contrast patterns. *Journal of Biomedical Informatics*, 66:19–31, 2017. DOI: 10.1016/j.jbi.2016.12.010 91, 99

[89] Tao Gu, Liang Wang, Hanhua Chen, Guimei Liu, Xianping Tao, and Jian Lu. Mining emerging sequential patterns for activity recognition in body sensor networks. In *International Conference on Mobile and Ubiquitous Systems: Computing, Networking, and Services*, pages 102–113, Springer, 2010. DOI: 10.1007/978-3-642-29154-8_9 94, 99

[90] Tao Gu, Zhanqing Wu, XianPing Tao, Hung Keng Pung, and Jian Lu. epSICAR: An emerging patterns based approach to sequential, interleaved and concurrent activity recognition. In *Proc. of IEEE International Conference on Pervasive Computing and Communications (PerCom)*, pages 1–9, 2009. DOI: 10.1109/percom.2009.4912776 94, 99

[91] Jiawei Han, Jian Pei, and Micheline Kamber. *Data Mining: Concepts and Techniques*. Elsevier, 2011. 44

[92] Jiawei Han, Jian Pei, Yiwen Yin, and Runying Mao. Mining frequent patterns without candidate generation: A frequent-pattern tree approach. *Data Mining and Knowledge Discovery*, 8(1):53–87, 2004. DOI: 10.1023/b:dami.0000005258.31418.83 62

[93] Franciso Herrera, Cristóbal José Carmona, Pedro González, and María José Del Jesus. An overview on subgroup discovery: Foundations and applications. *Knowledge and Information Systems*, 29(3):495–525, 2011. DOI: 10.1007/s10115-010-0356-2 8

[94] Victoria Hodge and Jim Austin. A survey of outlier detection methodologies. *Artificial Intelligence Review*, 22(2):85–126, 2004. DOI: 10.1023/b:aire.0000045502.10941.a9 44

[95] Robert C. Holte, Liane Acker, Bruce W. Porter, et al. Concept learning and the problem of small disjuncts. In *International Joint Conferences on Artificial Intelligence (IJCAI)*, vol. 89, pages 813–818, 1989. 28, 33

[96] Oleg Iegorov, Vincent Leroy, Alexandre Termier, Jean-François Méhaut, and Miguel Santana. Data mining approach to temporal debugging of embedded streaming applications. In *Proc. of the 12th International Conference on Embedded Software*, pages 167–176, IEEE Press, 2015. DOI: 10.1109/emsoft.2015.7318272 98

[97] James Jaccard, Robert Turrisi, and Jim Jaccard. *Interaction Effects in Multiple Regression*, no. 72, Sage, 2003. DOI: 10.4135/9781412984522 68

[98] James Jaccard, Choi K. Wan, and Robert Turrisi. The detection and interpretation of interaction effects between continuous variables in multiple regression. *Multivariate Behavioral Research*, 25(4):467–478, 1990. DOI: 10.1207/s15327906mbr2504_4 68

[99] Ioannis Kavakiotis, George Tzanis, and Ioannis Vlahavas. Polyadenylation site prediction using polyA-iEP method. In *Polyadenylation*, pages 131–140, Springer, 2014. DOI: 10.1007/978-1-62703-971-0_11 97, 99

[100] Hyunjoong Kim, Wei-Yin Loh, Yu-Shan Shih, and Probal Chaudhuri. Visualizable and interpretable regression models with good prediction power. *IIE Transactions*, 39(6):565–579, 2007. DOI: 10.1080/07408170600897502 83

[101] Young Bun Kim, Jung Hun Oh, and Jean Gao. Emerging pattern based subspace clustering of microarray gene expression data using mixture models. In *Advances in Bioinformatics and its Applications*, pages 13–24, World Scientific, 2005. DOI: 10.1142/9789812702098_0002 98

[102] Sofia Kleisarchaki. Difference analysis in big data: Exploration, explanation, evolution. Ph.D. thesis, Université Grenoble Alpes, 2016. 98

[103] Willi Klösgen. Explora: A multipattern and multistrategy discovery assistant. In *Advances in Knowledge Discovery and Data Mining*, pages 249–271, American Association for Artificial Intelligence, 1996. 8

[104] Ludmila I. Kuncheva and Christopher J. Whitaker. Measures of diversity in classifier ensembles and their relationship with the ensemble accuracy. *Machine Learning*, 51(2):181–207, 2003. DOI: 10.1023/A:1022859003006 83

[105] Paul Labute. Binary QSAR: A new method for the determination of quantitative structure activity relationships. In *Biocomputing*, pages 444–455, World Scientific, 1999. DOI:
10.1142/9789814447300_0044 38

[106] Himabindu Lakkaraju, Stephen H. Bach, and Jure Leskovec. Interpretable decision sets:
A joint framework for description and prediction. In *Proc. of ACM SIGKDD International Conference on Knowledge Discovery and Data Mining*, pages 1675–1684, ACM,
2016. DOI: 10.1145/2939672.2939874 99

[107] Heon Gyu Lee, Kiyong Noh, Bum Ju Lee, Ho-Sun Shon, and Keun Ho Ryu. Cardio-
vascular disease diagnosis method by emerging patterns. In *Proc. of the 2nd International
Conference on Advanced Data Mining and Applications (ADMA)*, pages 819–826, 2006.
DOI: 10.1007/11811305_89 98, 99

[108] Hochang B. Lee and Constantine G. Lyketsos. Depression in Alzheimer's disease:
Heterogeneity and related issues. *Biological Psychiatry*, 54(3):353–362, 2003. DOI:
10.1016/s0006-3223(03)00543-2 83

[109] Jong Bum Lee, Minghao Piao, and Keun Ho Ryu. Incremental emerging patterns mining
for identifying safe and non-safe power load lines. In *IEEE International Conference
on Computer and Information Technology (CIT)*, pages 1424–1429, IEEE, 2010. DOI:
10.1109/cit.2010.255 97

[110] Alban Lepailleur, Guillaume Poezevara, and Ronan Bureau. Automated detection of
structural alerts (chemical fragments) in (eco) toxicology. *Computational and Structural
Biotechnology Journal*, 5(6), 2013. DOI: 10.5936/csbj.201302013 98

[111] Gang Li, Rob Law, Huy Quan Vu, Jia Rong, and Xinyuan Roy Zhao. Identifying emerging hotel preferences using emerging pattern mining technique. *Tourism Management*,
46:311–321, 2015. DOI: 10.1016/j.tourman.2014.06.015 97

[112] Jinyan Li, Guozhu Dong, and Kotagiri Ramamohanarao. Making use of the most expressive jumping emerging patterns for classification. In *Proc. of Pacific-Asia Conference on Knowledge Discovery and Data Mining (PAKDD)*, pages 220–232, 2000. DOI:
10.1007/3-540-45571-x_29 17

[113] Jinyan Li, Guozhu Dong, and Kotagiri Ramamohanarao. Making use of the most expressive jumping emerging patterns for classification. *Knowledge and Information Systems*,
3(2):131–145, 2001. DOI: 10.1007/3-540-45571-x_29 35

[114] Jinyan Li, Guozhu Dong, Kotagiri Ramamohanarao, and Limsoon Wong. DeEPs: A
new instance-based lazy discovery and classification system. *Machine Learning*, 54(2):99–
124, 2004. DOI: 10.1023/b:mach.0000011804.08528.7d 34, 36

[115] Jinyan Li, Haiquan Li, Limsoon Wong, Jian Pei, and Guozhu Dong. Minimum description length principle: Generators are preferable to closed patterns. In *AAAI Conference on Artificial Intelligence*, pages 409–414, 2006. 13

[116] Jinyan Li, Guimei Liu, and Limsoon Wong. Mining statistically important equivalence classes and delta-discriminative emerging patterns. In *Proc. of ACM Conference on Knowledge Discovery and Data Mining (KDD)*, pages 430–439, 2007. DOI: 10.1145/1281192.1281240 25, 55

[117] Jinyan Li, Huiqing Liu, James R. Downing, Allen Eng-Juh Yeoh, and Limsoon Wong. Simple rules underlying gene expression profiles of more than six subtypes of acute lymphoblastic leukemia (ALL) patients. *Bioinformatics*, 19(1):71–78, 2003. DOI: 10.1093/bioinformatics/19.1.71 30, 89, 99

[118] Jinyan Li, Kotagiri Ramamohanarao, and Guozhu Dong. Combining the strength of pattern frequency and distance for classification. In *Pacific-Asia Conference on Knowledge Discovery and Data Mining*, pages 455–466, Springer, 2001. DOI: 10.1007/3-540-45357-1_48 35, 95

[119] Jinyan Li and Limsoon Wong. Identifying good diagnostic gene groups from gene expression profiles using the concept of emerging patterns. *Bioinformatics*, 18(10):1406–1407, 2002. DOI: 10.1093/bioinformatics/18.10.1406 72

[120] Wenmin Li, Jiawei Han, and Jian Pei. CMAR: Accurate and efficient classification based on multiple class-association rules. In *Proc. of IEEE International Conference on Data Mining (ICDM)*, pages 369–376, 2001. DOI: 10.1109/icdm.2001.989541 35

[121] Kah Leong Lim, Prasanna R. Kolatkar, Kwok Peng Ng, Chee Hoe Ng, and Catherine J. Pallen. Interconversion of the kinetic identities of the tandem catalytic domains of receptor-like protein-tyrosine phosphatase PTPα by two point mutations is synergistic and substrate-dependent. *Journal of Biological Chemistry*, 273(44):28986–28993, 1998. DOI: 10.1074/jbc.273.44.28986 25

[122] Bing Liu, Wynne Hsu, and Yiming Ma. Integrating classification and association rule mining. In *Proc. of ACM Conference on Knowledge Discovery and Data Mining (KDD)*, pages 80–86, 1998. 35

[123] Guimei Liu, Haojun Zhang, Mengling Feng, Limsoon Wong, and See-Kiong Ng. Supporting exploratory hypothesis testing and analysis. *ACM Transactions on Knowledge Discovery from Data (TKDD)*, 9(4):31, 2015. DOI: 10.1145/2701430 24, 97

[124] Lu Liu, Lei Duan, Hao Yang, Jyrki Nummenmaa, Guozhu Dong, and Pan Qin. Mining distinguishing customer focus sets for online shopping decision support. In *International*

Conference on Advanced Data Mining and Applications, pages 50–64, Springer, 2016. DOI: 10.1007/978-3-319-49586-6_4 97

[125] Qingbao Liu and Guozhu Dong. A contrast pattern based clustering quality index for categorical data. In *Proc. of IEEE International Conference on Data Mining (ICDM)*, pages 860–865, 2009. DOI: 10.1109/ICDM.2009.105 51, 52, 53, 54, 55, 56

[126] Qingbao Liu and Guozhu Dong. CPCQ: Contrast pattern based clustering quality index for categorical data. *Pattern Recognition*, 45(4):1739–1748, 2012. DOI: 10.1016/j.patcog.2011.10.007 51, 52, 53, 54, 55, 56

[127] Quanzhong Liu, Jiangning Song, and Jinyan Li. Using contrast patterns between true complexes and random subgraphs in PPI networks to predict unknown protein complexes. *Scientific Reports*, 6, 2016. DOI: 10.1038/srep21223 92, 99

[128] Xiaoqing Liu, Jun Wu, Feiyang Gu, Jie Wang, and Zengyou He. Discriminative pattern mining and its applications in bioinformatics. *Briefings in Bioinformatics*, 16(5):884–900, 2014. DOI: 10.1093/bib/bbu042 8

[129] Yong Liu, Jianzhong Li, and Jinghua Zhu. A novel graph classification approach based on frequent closed emerging patterns. *Journal of Computer Research and Development*, 44(7):1169–1176, 2007. DOI: 10.1360/crad20070711 95

[130] Elsa Loekito and James Bailey. Fast mining of high dimensional expressive contrast patterns using zero-suppressed binary decision diagrams. In *Proc. of the 12th ACM SIGKDD International Conference on Knowledge Discovery and Data Mining*, pages 307–316, 2006. DOI: 10.1145/1150402.1150438 8

[131] Octavio Loyola-González, José Fco Martínez-Trinidad, Jesús Ariel Carrasco-Ochoa, and Milton García-Borroto. Study of the impact of resampling methods for contrast pattern based classifiers in imbalanced databases. *Neurocomputing*, 175:935–947, 2016. DOI: 10.1016/j.neucom.2015.04.120 33, 36

[132] Octavio Loyola-González, José Fco Martínez-Trinidad, Jesús Ariel Carrasco-Ochoa, and Milton García-Borroto. A novel contrast pattern selection method for class imbalance problems. In *Mexican Conference on Pattern Recognition*, pages 42–52, Springer, 2017. DOI: 10.1007/978-3-319-59226-8_5 96

[133] Octavio Loyola-González, Miguel Angel Medina-Pérez, Dayton Hernández-Tamayo, Raúl Monroy, Jesús Ariel Carrasco-Ochoa, and Milton García-Borroto. A pattern-based approach for detecting pneumatic failures on temporary immersion bioreactors. *Sensors*, 19(2):414, 2019. DOI: 10.3390/s19020414 96

[134] Octavio Loyola-González, Miguel Angel Medina-Pérez, José Fco Martínez-Trinidad, Jesús Ariel Carrasco-Ochoa, Raúl Monroy, and Milton García-Borroto. PBC4cip: A new contrast pattern-based classifier for class imbalance problems. *Knowledge-Based Systems*, 115:100–109, 2017. DOI: 10.1016/j.knosys.2016.10.018 36, 96

[135] Sylvain Lozano, Guillaume Poezevara, Marie-Pierre Halm-Lemeille, Elodie Lescot-Fontaine, Alban Lepailleur, Ryan Bissell-Siders, Bruno Crémilleux, Sylvain Rault, Bertrand Cuissart, and Ronan Bureau. Introduction of jumping fragments in combination with QSARs for the assessment of classification in ecotoxicology. *Journal of Chemical Information and Modeling*, 50(8):1330–1339, 2010. DOI: 10.1021/ci100092x 98

[136] Hadi Tabatabaee Malazi and Mohammad Davari. Combining emerging patterns with random forest for complex activity recognition in smart homes. *Applied Intelligence*, 48(2):315–330, 2018. DOI: 10.1007/s10489-017-0976-2 94, 99

[137] Shihong Mao and Guozhu Dong. Discovery of highly differentiative gene groups from microarray gene expression data using the gene club approach. *Journal of Bioinformatics and Computational Biology*, 3(6):1263–1280, 2005. DOI: 10.1142/s0219720005001545 70, 75

[138] Shihong Mao and Guozhu Dong. *Towards Mining Optimal Emerging Patterns Amidst 1000s of Genes*. Chapter in *Contrast Data Mining: Concepts, Algorithms and Applications*, Guozhu Dong and James Bailey, Eds., Chapman & Hall/CRC, Data Mining and Knowledge Discovery Series, 2012. 75

[139] Shihong Mao, Charles Wang, and Guozhu Dong. Evaluation of inter-laboratory and cross-platform concordance of DNA microarrays through discriminating genes and classifier transferability. *Journal of Bioinformatics and Computational Biology*, 7(01):157–173, 2009. DOI: 10.1142/s0219720009004011 98

[140] Jean-Philippe Métivier, Bertrand Cuissart, Ronan Bureau, and Alban Lepailleur. The pharmacophore network: A computational method for exploring structure—activity relationships from a large chemical data set. *Journal of Medicinal Chemistry*, 61(8):3551–3564, 2018. DOI: 10.1021/acs.jmedchem.7b01890 88, 98

[141] Jean-Philippe Métivier, Alban Lepailleur, Aleksey Buzmakov, Guillaume Poezevara, Bruno Crémilleux, Sergei O. Kuznetsov, J'erémie Le Goff, Amedeo Napoli, Ronan Bureau, and Bertrand Cuissart. Discovering structural alerts for mutagenicity using stable emerging molecular patterns. *Journal of Chemical Information and Modeling*, 55(5):925–940, 2015. DOI: 10.1021/ci500611v 25, 89, 99

[142] Dean Mobbs, Cindy C. Hagan, Tim Dalgleish, Brian Silston, and Charlotte Prévost. The ecology of human fear: Survival optimization and the nervous system. *Frontiers in Neuroscience*, 9:55, 2015. DOI: 10.3389/fnins.2015.00055 2

[143] MOE (Molecular Operating EnVironment). Chemical Computing Group Inc., Montreal, Quebec, Canada H3B 3X3. 39

[144] David S. Moore, George P. McCabe, and Bruce A. Craig. *Introduction to the Practice of Statistics*. W H Freeman New York, 2009. DOI: 10.2307/1269120 24

[145] Hiroyuki Morita and Yukinobu Hamuro. A classification model using emerging patterns incorporating item taxonomy. In *Recent Progress in Data Engineering and Internet Technology*, pages 187–192, Springer, 2013. DOI: 10.1007/978-3-642-28807-4_26 36

[146] Takanobu Nakahara. Use of personal color and purchasing patterns for distinguishing fashion sensitivity. In *International Conference on Social Computing and Social Media*, pages 258–267, Springer, 2018. DOI: 10.1007/978-3-319-91485-5_20 98

[147] Mike A. Nalls, Nathan Pankratz, Christina M. Lill, Chuong B. Do, Dena G. Hernandez, Mohamad Saad, Anita L. DeStefano, Eleanna Kara, Jose Bras, Manu Sharma, et al. Large-scale meta-analysis of genome-wide association data identifies six new risk loci for Parkinson's disease. *Nature Genetics*, 46(9):989, 2014. DOI: 10.1038/ng.3043 67

[148] Vigneshwaran Namasivayam, Disha Gupta-Ostermann, Jenny Balfer, Kathrin Heikamp, and Jurgen Bajorath. Prediction of compounds in different local structure-activity relationship environments using emerging chemical patterns. *Journal of Chemical Information and Modeling*, 54(5):1301–1310, 2014. DOI: 10.1021/ci500147b 88, 98

[149] Vigneshwaran Namasivayam, Ye Hu, Jenny Balfer, and Jurgen Bajorath. Classification of compounds with distinct or overlapping multi-target activities and diverse molecular mechanisms using emerging chemical patterns. *Journal of Chemical Information and Modeling*, 53(6):1272–1281, 2013. DOI: 10.1021/ci400186n 88, 98

[150] Vigneshwaran Namasivayam, Preeti Iyer, and Jurgen Bajorath. Prediction of individual compounds forming activity cliffs using emerging chemical patterns. *Journal of Chemical Information and Modeling*, 53(12):3131–3139, 2013. DOI: 10.1021/ci400597d 88, 98

[151] Kerstin Neubarth and Darrell Conklin. Contrast pattern mining in folk music analysis. In *Computational Music Analysis*, pages 393–424, Springer, 2016. DOI: 10.1007/978-3-319-25931-4_15 90

[152] Mao Nishiguchi and Hiroyuki Morita. CAECP and CRPD: Classification by aggregating essential contrast patterns, and contrast ranked path diagrams. *Journal of Information and Knowledge Management*, 15(04), 2016. DOI: 10.1142/s0219649216500453 36

[153] Petra Kralj Novak, Nada Lavrac, and Geoffrey I. Webb. Supervised descriptive rule discovery: A unifying survey of contrast set, emerging pattern and subgroup mining. *Journal of Machine Learning Research*, 10:377–403, 2009. 8

[154] Megan Oaten, Richard J. Stevenson, and Trevor I. Case. Disgust as a disease-avoidance mechanism. *Psychological Bulletin*, 135(2):303, 2009. DOI: 10.1037/a0014823 2

[155] Katsuhiko Okada, Takahiro Azuma, Masakazu Nakamoto, and Yukinobu Hamuro. Stock performance after securities analyst's rating downgrades. *Transactions of the Japanese Society for Artificial Intelligence*, 27(6):355–364, 2012. DOI: 10.1527/tjsai.27.355 97

[156] Dijana Oreski, Irena Kedmenec, and Bozidar Klicek. Exploring capabilities of contrast mining application in SWOT analysis. *Proc. of the 8th MAC*, page 210, 2016. 97

[157] Stephan Pabinger, Andreas Dander, Maria Fischer, Rene Snajder, Michael Sperk, Mirjana Efremova, Birgit Krabichler, Michael R. Speicher, Johannes Zschocke, and Zlatko Trajanoski. A survey of tools for variant analysis of next-generation genome sequencing data. *Briefings in Bioinformatics*, 15(2):256–278, 2014. DOI: 10.1093/bib/bbs086 66, 68

[158] Xianchao Pan, Hu Mei, Sujun Qu, Shuheng Huang, Jiaying Sun, Li Yang, and Hua Chen. Prediction and characterization of P-glycoprotein substrates potentially bound to different sites by emerging chemical pattern and hierarchical cluster analysis. *International Journal of Pharmaceutics*, 502(1–2):61–69, 2016. DOI: 10.1016/j.ijpharm.2016.02.022 33

[159] Jin Hyoung Park, Heon Gyu Lee, Gyoyong Sohn, Jin-Ho Shin, and Keun Ho Ryu. Emerging pattern based classification for automated non-safe power line detection. In *Proc. of International Conference on Fuzzy Systems and Knowledge Discovery (FSKD)*, pages 169–173, 2009. DOI: 10.1109/fskd.2009.769 97

[160] Nicolas Pasquier, Yves Bastide, Rafik Taouil, and Lotfi Lakhal. Discovering frequent closed itemsets for association rules. In *Proc. of International Conference on Database Theory*, pages 398–416, 1999. DOI: 10.1007/3-540-49257-7_25 13

[161] Jun Pei and Min Zhang. A new classification algorithm based on emerging patterns. In *Proc. of the 6th International Asia Conference on Industrial Engineering and Management Innovation*, pages 57–65, Springer, 2016. DOI: 10.2991/978-94-6239-145-1_6 95

[162] Minghao Piao, Heon Gyu Lee, Gyoyong Sohn, Gouchol Pok, and Keun Ho Ryu. Emerging patterns based methodology for prediction of patients with myocardial ischemia. In *Proc. of International Conference on Fuzzy Systems and Knowledge Discovery (FSKD)*, pages 174–178, 2009. DOI: 10.1109/fskd.2009.638 33, 98, 99

[163] Guillaume Poezevara, Bertrand Cuissart, and Bruno Crémilleux. Discovering emerging graph patterns from chemicals. In *Proc. of International Symposium on Foundations of Intelligent Systems (ISMIS)*, pages 45–55, 2009. DOI: 10.1007/978-3-642-04125-9_8 98

[164] Guillaume Poezevara, Sylvain Lozano, Bertrand Cuissart, Ronan Bureau, Pierre Bureau, Vincent Croixmarie, Philippe Vayer, and Alban Lepailleur. A computational selection of metabolite biomarkers using emerging pattern mining: A case study in human hepatocellular carcinoma. *Journal of Proteome Research*, 16(6):2240–2249, 2017. DOI: 10.1021/acs.jproteome.7b00054 88, 99

[165] Adriana Prado, Marc Plantevit, Céline Robardet, and Jean-Francois Boulicaut. Mining graph topological patterns: Finding covariations among vertex descriptors. *IEEE Transactions on Knowledge and Data Engineering*, 25(9):2090–2104, 2013. DOI: 10.1109/tkde.2012.154 97

[166] Hemant Purohit, Guozhu Dong, Valerie Shalin, Krishnaprasad Thirunarayan, and Amit Sheth. Intent classification of short-text on social media. In *IEEE International Conference on Smart City/SocialCom/SustainCom (SmartCity)*, pages 222–228, 2015. DOI: 10.1109/smartcity.2015.75 97

[167] J. Ross Quinlan. Learning efficient classification procedures and their application to chess end games. In *Machine Learning*, pages 463–482, Springer, 1983. DOI: 10.1016/b978-0-08-051054-5.50019-4 2

[168] J. Ross Quinlan. *C4.5: Programs for Machine Learning*. Morgan Kaufmann, 1993. 35

[169] Murugesan Raju, Danlu Liu, Frederick W. Fraunfelder, and Chi-Ren Shyu. Discovering multifactorial associations with the development of age-related cataract using contrast mining. In *IEEE International Conference on Bioinformatics and Biomedicine (BIBM)*, pages 2297–2299, 2017. DOI: 10.1109/bibm.2017.8218033 98

[170] Kotagiri Ramamohanarao and Hongjian Fan. Patterns based classifiers. *World Wide Web*, 10(1):71–83, 2007. DOI: 10.1007/s11280-006-0012-7 36

[171] Anna Rampini, Gloria Bordogna, Paola Carrara, Monica Pepe, Massimo Antoninetti, Alessandro Mondini, and Paola Reichenbach. Modelling landslides' susceptibility by fuzzy emerging patterns. In *Landslide Science and Practice*, pages 363–370, Springer, 2013. DOI: 10.1007/978-3-642-31325-7_48 97

[172] Jenna M. Reps, Uwe Aickelin, and Richard B. Hubbard. Refining adverse drug reaction signals by incorporating interaction variables identified using emergent pattern mining. *Computers in Biology and Medicine*, 2015. DOI: 10.2139/ssrn.2822199 68, 90, 99

[173] Marylyn D. Ritchie, Lance W. Hahn, and Jason H. Moore. Power of multifactor dimensionality reduction for detecting gene-gene interactions in the presence of genotyping error, missing data, phenocopy, and genetic heterogeneity. *Genetic Epidemiology: The Official Publication of the International Genetic Epidemiology Society*, 24(2):150–157, 2003. DOI: 10.1002/gepi.10218 69

[174] Marylyn D. Ritchie, Lance W. Hahn, Nady Roodi, L. Renee Bailey, William D. Dupont, Fritz F. Parl, and Jason H. Moore. Multifactor-dimensionality reduction reveals high-order interactions among estrogen-metabolism genes in sporadic breast cancer. *The American Journal of Human Genetics*, 69(1):138–147, 2001. DOI: 10.1086/321276 68

[175] Guillaume A. Rousselet, Cyril R. Pernet, and Rand R. Wilcox. Beyond differences in means: Robust graphical methods to compare two groups in neuroscience. *European Journal of Neuroscience*, 46(2):1738–1748, 2017. DOI: 10.1111/ejn.13610 2, 24

[176] Khalid E. K. Saeed, Heon Gyu Lee, Wun-Jae Kim, Eun Jong Cha, and Keun Ho Ryu. Using emerging subsequence in classifying protein structural class. In *Proc. of International Conference on Fuzzy Systems and Knowledge Discovery (FSKD)*, pages 349–353, 2009. DOI: 10.1109/fskd.2009.752 98, 99

[177] David Savage. Detection of illicit behaviours and mining for contrast patterns. Ph.D. thesis, RMIT, Australia, 2017. 98

[178] Chinar C. Shah, Xingquan Zhu, Taghi M. Khoshgoftaar, and Justin Beyer. Contrast pattern mining with gap constraints for peptide folding prediction. In *FLAIRS Conference*, pages 95–100, 2008. 97, 99

[179] Jingbo Shang, Meng Jiang, Wenzhu Tong, Jinfeng Xiao, Jian Peng, and Jiawei Han. DPPred: An effective prediction framework with concise discriminative patterns. *IEEE Transactions on Knowledge and Data Engineering*, 2017. DOI: 10.1109/tkde.2017.2757476 8, 25

[180] Claude E. Shannon. A mathematical theory of communication. *Bell System Technical Journal*, 27:623–656, 1948. 11

[181] Richard Sherhod, Valerie J. Gillet, Philip N. Judson, and Jonathan D. Vessey. Automating knowledge discovery for toxicity prediction using jumping emerging pattern mining. *Journal of Chemical Information and Modeling*, 52(11):3074–3087, 2012. DOI: 10.1021/ci300254w 89, 98

[182] Richard Sherhod, Philip N. Judson, Thierry Hanser, Jonathan D. Vessey, Samuel J. Webb, and Valerie J. Gillet. Emerging pattern mining to aid toxicological knowledge discovery. *Journal of Chemical Information and Modeling*, 54(7):1864–1879, 2014. DOI: 10.1021/ci5001828 98

[183] Nicholas Skapura and Guozhu Dong. Class distribution curve based discretization with application to wearable sensors and medical monitoring. *International Journal of Monitoring and Surveillance Technologies Research (IJMSTR)*, 5(4):23–37, 2017. DOI: 10.4018/ijmstr.2017100102 11

[184] Hee Seok Song, Jae kyeong Kim, and Soung Hie Kim. Mining the change of customer behavior in an internet shopping mall. *Expert Systems with Applications*, 21(3):157–168, 2001. DOI: 10.1016/s0957-4174(01)00037-9 97

[185] Arnaud Soulet, Bruno Crémilleux, and François Rioult. Condensed representation of emerging patterns. In *Pacific-Asia Conference on Knowledge Discovery and Data Mining*, pages 127–132, Springer, 2004. DOI: 10.1007/978-3-540-24775-3_16 25

[186] Arnaud Soulet and C. Hébert. Using emerging patterns from clusters to characterize social subgroups of patients affected by atherosclerosis. In *Proc. of Discovery Challenge Workshop co-located with ECML/PKDD*, 2004. 2

[187] Matt Spencer, Nicole Takahashi, Sounak Chakraborty, Judith Miles, and Chi-Ren Shyu. Heritable genotype contrast mining reveals novel gene associations specific to autism subgroups. *Journal of Biomedical Informatics*, 77:50–61, 2018. DOI: 10.1016/j.jbi.2017.11.016 8, 96

[188] Michael Steinbach, Haoyu Yu, Gang Fang, and Vipin Kumar. Using constraints to generate and explore higher order discriminative patterns. In *Proc. of Pacific-Asia Conference on Knowledge Discovery and Data Mining (PAKDD)*, pages 338–350, 2011. DOI: 10.1007/978-3-642-20841-6_28 8

[189] Qun Sun, Xiuzhen Zhang, and Kotagiri Ramamohanarao. Noise tolerance of ep-based classifiers. In *Australasian Joint Conference on Artificial Intelligence*, pages 796–806, Springer, 2003. DOI: 10.1007/978-3-540-24581-0_68 33

[190] Yanmin Sun, Andrew K. C. Wong, and Mohamed S. Kamel. Classification of imbalanced data: A review. *International Journal of Pattern Recognition and Artificial Intelligence*, 23(04):687–719, 2009. DOI: 10.1142/s0218001409007326 28

[191] Atsushi Takizawa. Classification and feature extraction of criminal occurrence points using CAEP with transductive clustering. *Procedia—Social and Behavioral Sciences*, 21:83–92, 2011. DOI: 10.1016/j.sbspro.2011.07.036 90

[192] Guanting Tang, Jian Pei, James Bailey, and Guozhu Dong. Mining multidimensional contextual outliers from categorical relational data. *Intelligent Data Analysis*, 19(5):1171–1192, 2015. DOI: 10.3233/ida-150764 50

[193] Vahid Taslimitehrani and Guozhu Dong. A new CPXR based logistic regression method and clinical prognostic modeling results using the method on traumatic brain injury. In *IEEE International Conference on BioInformatics and BioEngineering (BIBE)*, pages 283–290, 2014. DOI: 10.1109/bibe.2014.16 84, 99

[194] Vahid Taslimitehrani, Guozhu Dong, Naveen L. Pereira, Maryam Panahiazar, and Jyotishman Pathak. Developing EHR-driven heart failure risk prediction models using CPXR(Log) with the probabilistic loss function. *Journal of Biomedical Informatics*, 60:260–269, 2016. DOI: 10.1016/j.jbi.2016.01.009 84, 99

[195] Mahbod Tavallaee, Ebrahim Bagheri, Wei Lu, and Ali A. Ghorbani. A detailed analysis of the KDD CUP 99 data set. In *IEEE Symposium on Computational Intelligence for Security and Defense Applications (CISDA)*, pages 1–6, 2009. DOI: 10.1109/cisda.2009.5356528 48

[196] Paweł Terlecki. On the relation between jumping emerging patterns and rough set theory with application to data classification. In *Transactions on Rough Sets XII*, pages 236–338, Springer-Verlag, 2010. DOI: 10.1007/978-3-642-14467-7_13 8

[197] Pawel Terlecki and Krzysztof Walczak. Adaptive classification with jumping emerging patterns. In *Proc. of 3rd International Conference on Rough Sets and Knowledge Technology*, pages 39–46, 2008. DOI: 10.1007/978-3-540-79721-0_11 96

[198] Takahisa Toda. Hypergraph transversal computation with binary decision diagrams. In *International Symposium on Experimental Algorithms*, pages 91–102, Springer, 2013. DOI: 10.1007/978-3-642-38527-8_10 25

[199] George Tzanis, Ioannis Kavakiotis, and Ioannis Vlahavas. PolyA-iEP: A data mining method for the effective prediction of polyadenylation sites. *Expert Systems with Applications*, 38(10):12398–12408, 2011. DOI: 10.1016/j.eswa.2011.04.019 97, 99

[200] Joseph L. Usset, Rama Raghavan, Jonathan P. Tyrer, Valerie McGuire, Weiva Sieh, Penelope Webb, Jenny Chang-Claude, Anja Rudolph, Hoda Anton-Culver, Andrew Berchuck, et al. Assessment of multifactor gene—environment interactions and ovarian cancer risk: Candidate genes, obesity, and hormone-related risk factors. *Cancer Epidemiology and Prevention Biomarkers*, 25(5):780–790, 2016. DOI: 10.1158/1538-7445.am2015-4684 68

[201] Adriano Veloso, Wagner Meira Jr., and Mohammed J. Zaki. Lazy associative classification. In *6th International Conference on Data Mining (ICDM)*, pages 645–654, IEEE, 2006. DOI: 10.1109/icdm.2006.96 36

[202] Sebastián Ventura and José María Luna. *Supervised Descriptive Pattern Mining*. Springer, 2018. DOI: 10.1007/978-3-319-98140-6 8

[203] Winn Voravuthikunchai, Bruno Crémilleux, and Frédéric Jurie. Histograms of pattern sets for image classification and object recognition. In *Proc. of the IEEE Conference on Computer Vision and Pattern Recognition*, pages 224–231, 2014. DOI: 10.1109/cvpr.2014.36 97

[204] Jilles Vreeken, Matthijs Van Leeuwen, and Arno Siebes. KRIMP: Mining itemsets that compress. *Data Mining and Knowledge Discovery*, 23(1):169–214, 2011. DOI: 10.1007/s10618-010-0202-x 24

[205] Chunyu Wang, Yizhou Wang, and Alan L. Yuille. An approach to pose-based action recognition. In *Proc. of the IEEE Conference on Computer Vision and Pattern Recognition*, pages 915–922, 2013. DOI: 10.1109/cvpr.2013.123 91, 99

[206] Dawei Wang, Wei Ding, Henry Lo, Melissa Morabito, Ping Chen, Josue Salazar, and Tomasz Stepinski. Understanding the spatial distribution of crime based on its related variables using geospatial discriminative patterns. *Computers, Environment and Urban Systems*, 39:93–106, 2013. DOI: 10.1016/j.compenvurbsys.2013.01.008 8, 18

[207] Liang Wang, Yizhou Wang, Tingting Jiang, and Wen Gao. Instantly telling what happens in a video sequence using simple features. In *Proc. of IEEE Conference on Computer Vision and Pattern Recognition (CVPR)*, pages 3257–3264, 2011. DOI: 10.1109/cvpr.2011.5995377 93, 99

[208] Lusheng Wang, Hao Zhao, Guozhu Dong, and Jianping Li. On the complexity of finding emerging patterns. *Theoretical Computer Science*, 335(1):15–27, 2005. DOI: 10.1109/cmpsac.2004.1342691 25

[209] Xiaoting Wang. Trajectory mining in the context of the Internet of Things. Ph.D. thesis, University of Melbourne, 2017. 97

[210] Xiaoting Wang, Christopher Leckie, Hairuo Xie, and Tharshan Vaithianathan. Discovering the impact of urban traffic interventions using contrast mining on vehicle trajectory data. In *Pacific-Asia Conference on Knowledge Discovery and Data Mining*, pages 486–497, Springer, 2015. DOI: 10.1007/978-3-319-18038-0_38 97

[211] Zhou Wang, Hongjian Fan, and Kotagiri Ramamohanarao. Exploiting maximal emerging patterns for classification. In *Australasian Joint Conference on Artificial Intelligence*, pages 1062–1068, Springer, 2004. DOI: 10.1007/978-3-540-30549-1_102 95

[212] Geoffrey I. Webb, Shane M. Butler, and Douglas A. Newlands. On detecting differences between groups. In *Proc. of ACM Conference on Knowledge Discovery and Data Mining (KDD)*, pages 256–265, 2003. DOI: 10.1145/956755.956781 8

[213] Wei Wei, Jinjiu Li, Longbing Cao, Yuming Ou, and Jiahang Chen. Effective detection of sophisticated online banking fraud on extremely imbalanced data. *World Wide Web*, 16(4):449–475, 2013. DOI: 10.1007/s11280-012-0178-0 94

[214] Gary M. Weiss. The impact of small disjuncts on classifier learning. In *Data Mining*, pages 193–226, Springer, 2010. DOI: 10.1007/978-1-4419-1280-0_9 28

[215] Cheng-Hsiung Weng and Cheng-Kui Huang Tony. Observation of sales trends by mining emerging patterns in dynamic markets. *Applied Intelligence*, pages 1–15, 2018. DOI: 10.1007/s10489-018-1231-1 97

[216] Daniela M. Witten and Robert Tibshirani. A comparison of fold-change and the t-statistic for microarray data analysis. *Analysis*, 2007. 66

[217] Stefan Wrobel. An algorithm for multi-relational discovery of subgroups. In *Proc. of European Conference on Principles and Practice of Knowledge Discovery in Databases (PKDD)*, pages 78–87, 1997. DOI: 10.1007/3-540-63223-9_108 2

[218] Stefan Wrobel. Inductive logic programming for knowledge discovery in databases. In *Relational Data Mining*, pages 74–101, Springer, 2001. DOI: 10.1007/978-3-662-04599-2_4 8

[219] Baohua Wu, Lei Duan, Zhonghua Yu, Changjie Tang, and Jun Zhu. Birth defects detection algorithm based on emerging patterns. *Journal of Computer Applications*, 31(4):885–889, 2011. DOI: 10.3724/sp.j.1087.2011.00885 93, 99

[220] Dongkuan Xu and Yingjie Tian. A comprehensive survey of clustering algorithms. *Annals of Data Science*, 2(2):165–193, 2015. DOI: 10.1007/s40745-015-0040-1 58

[221] Rui Xu and Donald Wunsch. Survey of clustering algorithms. *IEEE Transactions on Neural Networks*, 16(3):645–678, 2005. DOI: 10.1109/tnn.2005.845141 51, 58

[222] Jingfeng Xue, Changzhen Hu, Kunsheng Wang, Rui Ma, and Jiaxin Zou. Metamorphic Malware detection technology based on aggregating emerging patterns. In *Proc. of International Conference Interaction Sciences*, pages 1293–1296, 2009. DOI: 10.1145/1655925.1656162 33

[223] S. Ben Yahia, Tarek Hamrouni, and E. Mephu Nguifo. Frequent closed itemset based algorithms: A thorough structural and analytical survey. *ACM SIGKDD Explorations Newsletter*, 8(1):93–104, 2006. DOI: 10.1145/1147234.1147248 13

[224] Yulan Yan, Chikara Hashimoto, Kentaro Torisawa, Takao Kawai, Jun'ichi Kazama, and Stijn De Saeger. Minimally supervised method for multilingual paraphrase extraction from definition sentences on the Web. In *Proc. of the Conference of the North American Chapter of the Association for Computational Linguistics: Human Language Technologies*, pages 63–73, 2013. 97

[225] Eng-Juh Yeoh, Mary E. Ross, Sheila A. Shurtleff, W. Kent Williams, Divyen Patel, Rami Mahfouz, Fred G. Behm, Susana C. Raimondi, Mary V. Relling, Anami Patel, et al. Classification, subtype discovery, and prediction of outcome in pediatric acute lymphoblastic leukemia by gene expression profiling. *Cancer Cell*, 1(2):133–143, 2002. DOI: 10.1016/s1535-6108(02)00032-6 89, 99

[226] Xiaoxin Yin and Jiawei Han. CPAR: Classification based on predictive association rules. In *Proc. of SIAM International Conference on Data Mining (SDM)*, 2003. DOI: 10.1137/1.9781611972733.40 35

[227] Kui Yu, Wei Ding, Dan A. Simovici, Hao Wang, Jian Pei, and Xindong Wu. Classification with streaming features: An emerging-pattern mining approach. *ACM Transactions on Knowledge Discovery from Data (TKDD)*, 9(4):30, 2015. DOI: 10.1145/2700409 96

[228] Kui Yu, Xindong Wu, Wei Ding, Hao Wang, and Hongliang Yao. Causal associative classification. In *11th IEEE International Conference on Data Mining (ICDM)*, pages 914–923, 2011. DOI: 10.1109/icdm.2011.30 33, 35

[229] Reza Zafarani and Huan Liu. Social computing data repository at ASU, http://soci alcomputing.asu.edu, 2009. 63

[230] Shaoyi Zhang, Kotagiri Ramamohanarao, and James C. Bezdek. EP-based robust weighting scheme for fuzzy SVMs. In *Proc. of the 21st Australasian Conference on Database Technologies*. 95

[231] Xiuzhen Zhang, Guozhu Dong, and Kotagiri Ramamohanarao. Information-based classification by aggregating emerging patterns. In *Proc. of Intelligent Data Engineering and Automated Learning (IDEAL)*, pages 48–53, 2000. DOI: 10.1007/3-540-44491-2_8 31, 35, 96

[232] Xiuzhen Zhang, Guozhu Dong, and Kotagiri Ramamohanarao. Building behaviour knowledge space to make classification decision. In *Pacific-Asia Conference on Knowledge Discovery and Data Mining*, pages 488–494, Springer, 2001. DOI: 10.1007/3-540-45357-1_51 95

[233] Zhigang Zheng, Wei Wei, Chunming Liu, Wei Cao, Longbing Cao, and Maninder Bhatia. An effective contrast sequential pattern mining approach to taxpayer behavior analysis. *World Wide Web*, 19(4):633–651, 2016. DOI: 10.1007/s11280-015-0350-4 96

[234] Yun Zhou, James Bailey, Ioanna Ioannou, Sudanthi Wijewickrema, Stephen O'Leary, and Gregor Kennedy. Pattern-based real-time feedback for a temporal bone simulator. In *Proc. of the 19th ACM Symposium on Virtual Reality Software and Technology*, pages 7–16, 2013. DOI: 10.1145/2503713.2503728 33, 93, 99

[235] Xiaojin Zhu. Semi-supervised learning literature survey. *Computer Science*, 2(3):4, University of Wisconsin-Madison, 2006. 40

Author's Biography

GUOZHU DONG

Dr. Guozhu Dong is a professor of Computer Science and Engineering, and a member at the Knoesis Center of Excellence, at Wright State University. He received a Ph.D. in Computer Science from the University of Southern California and a B.S. in Mathematics from Shandong University. Before joining Wright State University, he was a faculty member at the University of Melbourne. His research interests span data mining, machine learning, databases, data science, bioinformatics, and artificial intelligence. He co-authored the book *Sequence Data Mining*; co-edited two books, *Contrast Data Mining* and *Feature Engineering*, respectively; and authored the book *Exploiting the Power of Group Differences*. He is known for his pioneering work and sustained effort on emerging/contrast pattern mining and on the use of such patterns in problem solving. He has published hundreds of papers at major international conferences and in top-rate journals in the fields of data mining and databases. He received several best research paper awards at major data mining conferences. At Wright State University, he was recognized for Excellence in Research in his college. He has served on hundreds of program committees of international conferences, and he has chaired the program committees for several such conferences. He is a senior member of both ACM and IEEE.

Index

Printed in the United States
by Baker & Taylor Publisher Services